POLITECNICO DI BARI

Dottorato in Ingegneria dell'Informazione

Ciclo XXIV

Position Control with magnetic shape memory actuators

Relatori:

Prof. Biagio Turchiano

Prof. David Naso

Tesi di

Leonardo Riccardi

2012

ISBN: 978-1-291-07319-5

This thesis is the product of the collaboration with other researchers.

I would like to personally thank Benedikt Holz from LPA, whose great ideas about MSM actuators have given to me both the motivation for my work and the practical insights which are needed for design of such actuators.

I would like to thank prof. Hartmut Janocha from LPA, which has given me the possibility of working on the subject, and the possibility of meeting the very active german community working on MSM materials.

I would like to thank prof. David Naso and prof. Biagio Turchiano from Politecnico di Bari, for being the supervisors and references of my whole thesis work.

I am personally grateful to the students which have collaborated with me during their own thesis project, in particular Giuseppe Ciaccia and Michele Rosmarino, and to the colleagues at LPA and Politecnico di Bari, whose company and companionship has made the Ph.D period an unforgettable experience.

Index

Introduction

Magnetic Shape Memory (MSM) materials are new materials which exhibit a strain up to 5% when excited by a magnetic field. Since their discovery (1996), the research about those materials has been focused mainly on production issues, while the other aspects concerning the design of MSM-based actuators and the control of such actuators are receiving an increasing attention by the scientific community only in the past few years. The huge strain and the high maximum frequency make MSM materials the perfect candidates for positioning systems in the micrometer range. Unfortunately, the strong influence of temperature on the positioning performance discourages the industrial adoption of the MSM technology.

This thesis has been developed within a collaboration between the Politecnico di Bari and the Laboratory of Process Automation (LPA) of Saarland University in Saarbruecken, with the twofold goal of designing new actuator concepts and providing suitable control strategies for position control. The first activity has been pursued mainly by LPA. The second activity has been pursued by the Politecnico di Bari, and is the main focus of this thesis work. At the beginning, it has required the understanding of the basic effects of MSM materials, and also the development of a simple simulation model. Chapter 1 explains how the material works and presents the simulation model.

Chapter 2 is devoted to the analysis of MSM actuators. The basic concepts are introduced, and some prototypes for positioning are presented. Chapter 2 also emphasizes the main issues which are relevant from the control viewpoint: the presence of hysteresis and the fact that this nonlinearity is time-varying due to temperature variations.

Hysteresis can negatively affect the positioning control-loops, decreasing the performances or even leading to instability. For this reason, Chapter 3 briefly presents some models of the hysteresis phenomenon, such as the modified Prandtl-Ishlinskii model. Those models offer a compensator, i.e. an inverse model that can be used in feedback control-loops to compensate for the nonlinearity. The control of MSM actuators for positioning purposes belongs to the wider class of control of hysteretic dynamic systems.

Chapter 4 considers the control of MSM actuators for the tracking of constant references. In this case, it is possible to adopt simple controllers such as standard controllers. In Chapter 4 the main existing results about PI/PID control of hysteretic systems are presented, and a novel control approach is introduced. The final part of the chapter confirms the validity of the new approach by means of experimental results acquired on two different MSM prototypes.

Chapter 5 considers the control of MSM actuators for the tracking of arbitrary references, in the challenging case that the plant is influenced by bounded temperature disturbances. The proposed control strategy makes use of an adaptive modified Prandtl-Ishlinskii compensator. It is demonstrated that the proposed adaptive control can ensure a constant tracking performance in the presence of bounded temperature variations.

This thesis builds a bridge between the research areas focused on the MSM material development and the design of MSM-based actuators, and the application of such devices into effective positioning systems. Here, the reader will find ideas and suggestions about how to design a control system for MSM-actuators. Those hints can also be generalized and extended for any smart-material-based system.

1. MSM materials

In this chapter, the Magnetic Shape Memory (MSM) material is presented. The discussion is kept concise: it will focus on the main magneto-mechanical properties which are of direct interest for actuation. The details about the chemical structure of the MSM alloys (MSMAs) are avoided, as well as details about their production process.

Section 1.1 describes the physical effect of the MSMAs, and their reaction to the application of an external stress or field is shown.

Section1.2 illustrates the energy framework which is commonly used to derive models of smart-materials. In the first part the thermodynamic framework is recalled, and then it is applied to model MSM materials. Only the stress-field-strain relationship is obtained, while the behavior of the magnetization is neglected. The magnetization behavior becomes relevant when the material has to be analyzed as a force or strain sensor.

Section 1.3 compares MSM materials with other active materials such as piezoelectric ceramics, magnetostrictive and thermal shape memory alloys. The main quantities used for the comparison are introduced.

Section 1.4 closes the description of the material hinting some other useful effects of MSM alloys. For example, one effect is the change of the magnetic permeability with respect to the strain. A possible exploitation is in the development of strain sensors by measuring the permeability of the MSM element.

At present the research about MSM alloys is in progress. This holds true especially for their production process. It is not possible to rely on a "standard" process that offers MSM materials having common properties. Each producer has developed its own way to create those materials and, generally, a MSM sample can show different characteristics with respect to another sample, even within the same production series.

1.1. Physical effect

The most commonly used MSM material is a Nichel-Manganese-Gallium (NiMnGa) alloy which exhibits a strain of up to 6% when excited by the magnetic field [1], [2]. Since MSM materials are ferromagnetic compounds, we will refer to the magnetic field strength H as the input variable. The macroscopic strain ε is the output variable. In section 1.1.1 a panoramic view on the working principle of the material is given. The main issues are discussed and some keywords are introduced and defined. Section 1.1.2 discusses the relationships which are of interest from the actuation point of view.

1.1.1. An overview

MSM materials are produced at high temperature and then rapidly cooled down to obtain the necessary magneto-mechanical properties. When the alloy is at high temperature, it is said to be in the *austenite* phase. The microscopic structure of the crystal in the austenite phase has a cubic geometry, as can be seen in Figure 1.1 (left). The cooling procedure brings the alloy in a phase called *martensite* phase. In this phase, the unit cells of the crystal have a non-cubic, *tetragonal* structure, that is characterized by a short axis and a long axis. There appear different areas in the MSM crystal, each one composed of unit cells that have a specific orientation of their short axis. These areas are called *twin variants*, denoted with V_1, V_2 and V_3 in Figure 1.1. Each variant has a short axis and a long axis, respectively c and a. The short axis is also called *easy axis* since the magnetization elong its direction requires less energy with respect to the magnetization along the long, so-called *hard axis*. The twin variants generally coexist in the MSM crystal, unless external excitations such as a magnetic field H or an external stress σ force the variants to rearrange, favoring one despite of the others.

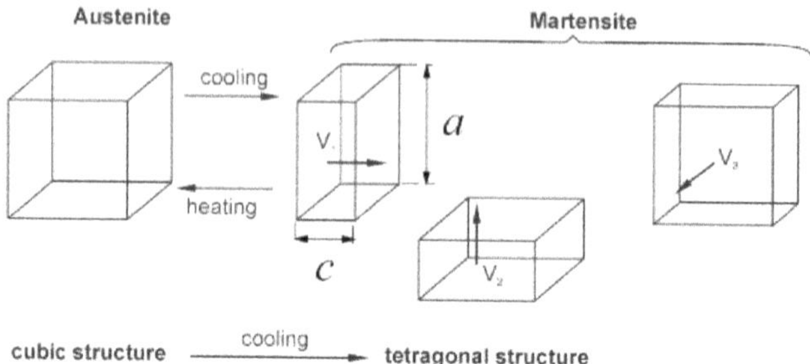

Figure 1.1 - Crystalline structure of the MSM alloy in austenite and martensite

To understand the basic working principle of MSM material, refer to Figure 1.2. Only two variants are of interest if the actuation is studied in two dimensions. Let us denote with y the direction where the magnetic field H applies, and with x the direction of the deformation of the material. The easy axes of the variants and their orientations are indicated by the small horizontal lines.

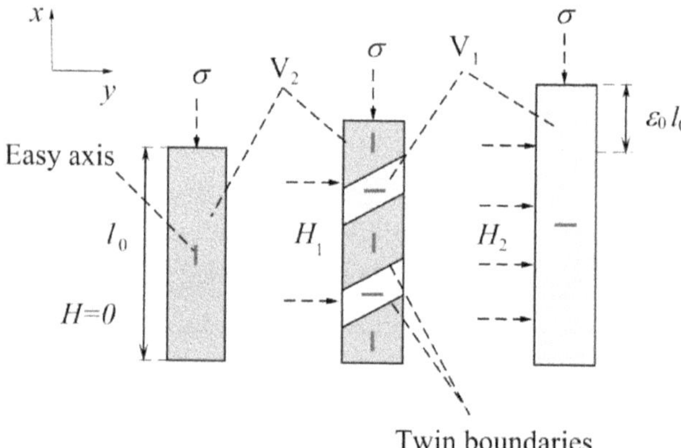

Figure 1.2 – Illustration of the MSM effect: from left to right the influence of the applied magnetic field H can be appreciated in the growing of the field-preferred variant V_1

Normally, the martensitic phase is ferromagnetic and internally twinned, i.e. the MSMA is formed by a series of parallel layers composed by alternating (twin)

variants which are separated by borders called *twin boundaries*. The situation is shown in Figure 1.2 in the central drawing. The presence of the variants is what makes the actuation mechanism possible. Let us assume that the crystal starts in a status where only the variant V_2 is present (Figure 1.2, left). This can be obtained by compressing the crystal with an external stress σ. The application of an external compressive stress favors the presence of that variant which has its short axis parallel to the direction of the stress, i.e. V_2. Let us denote the initial length (in complete contraction) with l_0. If the stress is acting and the field is zero, the material will remain in the contracted status. If the field strength is increased, the variant which has the easy axis parallel to the field direction, V_1, starts growing. This is illustrated in Figure 1.2, center. The growth of the variant is due to the movement of the twin boundaries. If the field strength increases up to a certain level, then the variant V_1 grows until the complete volume of the MSM crystal is composed by just V_1. This situation is illustrated in Figure 1.2, right. It corresponds to the situation where the MSM element has its maximum deformation, i.e. it is completely elongated.

The maximum deformation depends on the rotation of the unit cells. When all of them are rotated ($V_2 \rightarrow V_1$), the deformation is maximal. The maximum crystallographic strain ε_0 is defined as

$$\varepsilon_0 = 1 - \frac{c}{a}, \tag{1.1}$$

and it expresses the maximum deformation that can be obtained from a complete rotation of the unit cells. The maximum elongation Δl_{MAX} can be calculated by

$$\Delta l_{MAX} = \varepsilon_0 l_0, \tag{1.2}$$

as pointed out in the right part of Figure 1.2.

If the field is removed, the external stress will bring the crystal again in the contracted status by means of growth of the variant V_2. The process of rearrangement of the variants ($V_2 \rightarrow V_1$) is called *magnetically induced reorientation* (MIR) and is the basic effect of MSM material [2]. It is possible because the MSM material fulfills three important properties at the same time:

1. it has a non-cubic crystalline structure;
2. the crystal has a *high magnetic anisotropy*;
3. the variants are connected by highly mobile twin boundaries.

The property (1) is necessary since without a non-cubic structure there will be no crystallographic strain, $\varepsilon_0 = 0$. The non-cubic structure is responsible for the presence of a non-uniform distribution of the magnetic properties in the 3D space. Property (1) is however not sufficient to have strain. An example is given by some soft magnetic materials. They also exhibit a non-cubic structure and thus some difference in the magnetic properties along the 3D spatial directions, which is expressed by the presence of little magnetic domains with magnetic momenta that are differently oriented. If a magnetic field is applied to soft magnetic materials, it will force the momenta of all domains to rotate and align parallel to the magnetic field. However, there is no relevant macroscopic strain.

The property (2) involves the concept of *magnetocrystalline anisotropy*. This is an inherent property of the material and concerns the change of the energy in the crystal as the direction of magnetization is varied. Refer to Figure 1.1. The easy axis *c* of the unit cell is also the direction where the magnetization is "easier" in the sense that the process requires less energy. The *anisotropy energy* quantifies the increase in energy required to magnetize the unit cell along another direction. The *a* axis of the unit cells is called the *hard axis* since the magnetization along *a* is "harder", requiring much more energy. The maximum anisotropy energy is thus given when the crystal is magnetized along the *a* direction [1]. This quantity is called the *anisotropy constant*, and is a parameter of the MSM material. Having a high magnetic anisotropy means that the anisotropy constant is big, i.e. there is a big energy difference when magnetizing along the *c* axis or the *a* axis. Materials that have a big anisotropy constant are called *hard magnetic* materials. In this case, under the application of a magnetic field, the rotation of the unit cells is favored with respect to the rotation of their magnetic momenta only. In other words, thanks to the big anisotropy energy the unit cells rotate under magnetic field and thus the strain ε_0 is possible.

However, the fact that the rotation is favored and that the strain is possible from a crystallographic viewpoint are also not sufficient conditions for the macroscopic strain to happen. Condition (3) must be satisfied for practical reasons. If the twin boundaries are not highly mobile, the strain would occur only under very intense magnetic fields, making the use of the MSM effect practically impossible. As will be detailed in the next pages, the mobility of the twin boundaries is expressed by a characteristic parameter of the material, called *twinning stress*. It can be thought as an internal friction among the variants, opposing any rearrangement and hindering the MIR.

Thanks to (1)-(3), MSM materials exhibit the MIR. There is a couple of details that any user of MSM materials should keep in mind. The first concerns the mechanism of the contraction. Refer to Figure 1.2. The magnetic field is applied along *y*. The MIR happens because V_1 grows, since it has the easy axis parallel to the field. If the field is now rotated of 180 degrees (negative values), V_1 will still grow. In other words, a change in sign of the magnetic field does not produce contraction. To obtain contraction and thus favor the growth of V_2, the magnetic field should rotate of 90 degrees. This is a fundamental issue for the design of a MSM actuator. At present, a common way is to use a spring which applies a mechanical stress σ on the MSM element, and pushes it back to the contracted position.

The second detail concerns the influence of temperature on the actuation properties of the MSM element. Refer to Figure 1.1. A heating of the material brings it back to the austenite phase and restores the cubic structure of the unit cells. The property (1) becomes unverified; moreover, a cubic structure is magnetically isotropic, so also (2) is unverified; finally, twin boundaries between the variants will not exist if there are no variants... The magnetic actuation capability of the material is strongly affected by an increase of temperature over the value which marks the phase transition from martensite to austenite (around 60 °C).

It has to be noted that a restoring of the cubic structure is an intrinsic deformation process; actually, this is the effect exploited by common thermal shape memory alloys. However, the thermal shape change is not the main working principle of MSM material, and for this reason in this work it will not be treated.

1.1.2. Stress, magnetic field and strain

From the actuation point of view, the variables of interest are the stress σ, the magnetic field H and the strain ε, which is the output quantity. This section attempts at giving a qualitative idea of how those variables interact and what is their macroscopic result.

Figure 1.3 shows again a MSM element excited by H under the action of the stress σ. For simplicity, only two areas of variants are depicted.

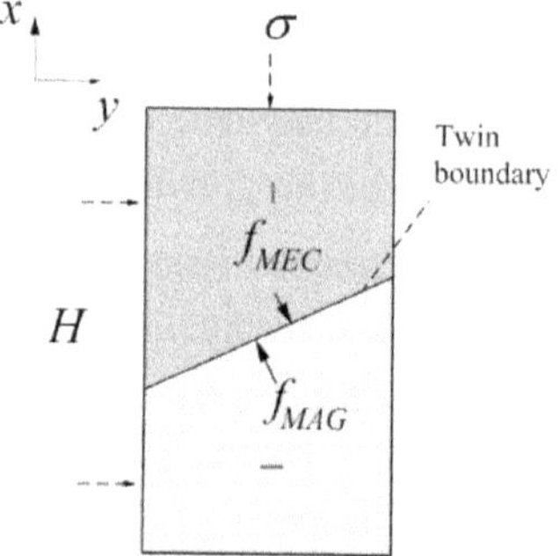

Figure 1.3 – Mechanical and magnetic microscopic forces acting on the twin boundaries

The application of σ generates the so-called microscopic (mechanical) driving force $f_{MEC} = f_{MEC}(\sigma)$, that acts on the twin boundaries in order to favor the growth of V_2. The magnetic field H generates the so-called microscopic (magnetic) driving force $f_{MAG} = f_{MAG}(H)$, which acts on the boundary in order to favor the growth of V_1. The movement of the boundary, and the corresponding growth of one specific variant, depend on the equilibrium between these microscopic forces. The resultant force has to overcome the twinning stress, acting as a friction on both directions, to allow starting a movement. The microscopic forces relate with the macroscopic stresses in this way [3]:

$$f_{MEC}(\sigma) = \varepsilon_0 \sigma , \tag{1.3}$$

$$f_{MAG}(H) = \varepsilon_0 \sigma_{MAG}(H) , \tag{1.4}$$

where $\sigma_{MAG}(H)$ denotes the *magnetic field induced stress* (MFIS). Its mathematical relationship with H will be defined later on.

Figure 1.3 suggests that the explanation of the MIR can be simplified to an equilibrium of forces acting on the boundaries. The application of a magnetic field, or the application of a mechanical stress have an almost equivalent effect on the boundary, since both produces some acting forces [3]. Let us discuss now the effect of the mechanical stress σ on the strain ε. The typical curve is depicted in Figure 1.4.

Let us assume that the MSM element starts in the contracted position, i.e. $\varepsilon = 0$, and that the initial value of the stress is zero, $\sigma = 0$. The only variant is V_2. This situation is represented by point *A* in Figure 1.4. The stress is taken positive when it pulls the MSM element towards elongation, and negative when it pushes the element back to contraction. When the stress increases, the MSM element shows an elastic deformation, up to a certain stress value σ_{TW} which is the twinning stress. Point *B* represents the critical point: in fact, when the stress $\sigma > \sigma_{TW}$, the variants rearrange from V_2 to V_1. The process ends in point *C*, where only the variant V_1 is present. If the stress continues increasing, then the MSM element shows again just an elastic deformation.

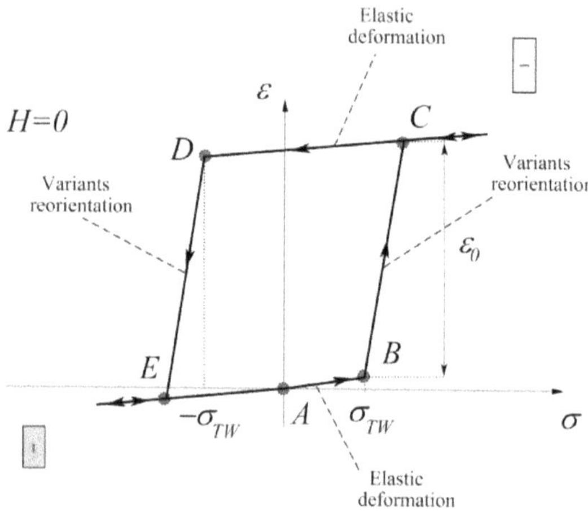

Figure 1.4 – Typical stress-strain curve of MSM materials

The stress is then reduced down to a value $-\sigma_{TW}$ described by point *D*. From *C* to *D* there is only an elastic deformation. When $\sigma < -\sigma_{TW}$ the reorientation from V_1 to V_2 starts. As soon as the variants are completely turned to V_2, i.e. the point *E* is reached, a range of elastic deformation starts.

We now discuss the effect of a magnetic field H, applied along the *y* direction (see Figure 1.3), on the strain ε. Since *H* generates a MFIS $\sigma_{MAG}(H)$, it is easier to present the field-strain relationship in terms of a $\sigma_{MAG}(H)$-ε curve. The typical characteristic is depicted in Figure 1.5. Let us assume, as before, that the material starts in the contracted position, i.e. at point *A* of Figure 1.5. Let us discuss the case where the external stress is zero. If *H* increases, then $\sigma_{MAG}(H)$ increases, but the MSM element will remain in the same status until $\sigma_{MAG} = \sigma_{TW}$, i.e., point *B*. As soon as $\sigma_{MAG} > \sigma_{TW}$, the variants reorientation starts and ends in point *C*. Any increase of the field will have no effect on the strain, which is at its maximum value $\varepsilon = \varepsilon_0$.

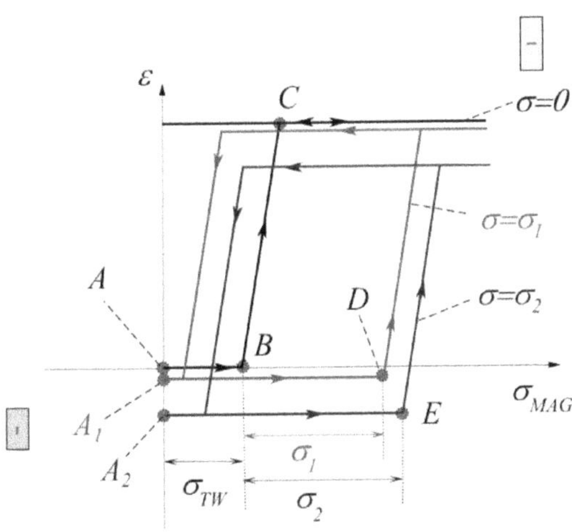

Figure 1.5 - A typical field-strain characteristic of MSM material, at different constant compressive (negative) stress values $\sigma = 0$, $\sigma = \sigma_1 < 0$ and $\sigma = \sigma_2 < \sigma_1$

However, if the field is now reduced, the strain will remain because of the shape memory effect of the material [4]. Moreover, if the field changes its sign, nothing will happen to the strain, since the MFIS only depends on the direction of the field. In other words, without any external stress and without the possibility to rotate the field from the *y* direction to the *x* direction of Figure 1.3, the material is not capable of actuation.

Let us now consider the case where an external constant compressive (and thus negative) stress $\sigma_1 < 0$ is applied on the MSM material. The initial contracted position is represented now by point A_1 in Figure 1.5. The difference in the ε coordinate of A and A_1 is due to the presence of the stress, which produces a certain amount of purely elastic strain depending on the Young modulus of the

MSM material. The MFIS $\sigma_{MAG}(H)$ must now overcome the joint contribution of σ_{TW} and σ_1 in order to start variant reorientation. The process, in fact, starts at point *D*, characterized by an abscissa value that is $|\sigma_1|+\sigma_{TW}$. It is worth noting that in this case the contracted status can be restored thanks to the presence of σ_1. Not any value of such a stress produces contraction, and this depends on the particular σ_{TW} of the MSM element that is considered. However, the condition $|\sigma_1| > \sigma_{TW}$ must be satisfied as a necessary condition to have shape restoring. If the field goes to negative values, the produced $\sigma_{MAG}(H)$ remains positive, and thus the same curves Figure 1.5 take place. This explains the typical "butterfly" characteristic of MSM materials that appears when driving them with a bipolar field.

If a constant stress $|\sigma_2| > |\sigma_1|$ is applied, the previous discussion still holds, with a different curve, a different starting point A_2 due to the elastic strain, and a different value of MFIS which starts the variants reorientation, i.e., $\sigma_{MAG}^{(E)} = +|\sigma_2| + \sigma_{TW}$. Obviously, it is always assumed that $|\sigma_2| > \sigma_{TW}$.

To summarize the discussion of this section, it is convenient to emphasize the following points:

- The output strain ε of the MSM material depends on the balance between the MFIS $\sigma_{MAG}(H)$, the mechanical stress σ and the twinning stress σ_{TW}.
- In order to have shape restoring and thus a reversible actuation mechanism, the MSM element has to be loaded by a stress σ, eventually constant, that overcomes the twinning stress, $|\sigma| > \sigma_{TW}$; or, in alternative, the magnetic field should be able to rotate its direction (complicated magnetic circuit design).

- In order for the MSM element to elongate magnetically, the MFIS has to overcome the twinning stress plus the external stress, so $\sigma_{MAG}(H) > \sigma_{TW} + |\sigma|$.
- The overall strain ε is composed by two parts. The first will be called twinning strain ε_{TW} and refers to the strain obtained by the variants reorientation. The maximum value of ε_{TW} is ε_0. The second part will be called the elastic strain ε_{EL}, and refers to the strain obtained by elastic deformation of the MSM element due to the external stress. Since the main phenomenon of MSM materials is the variant reorientation, $\varepsilon_{EL} << \varepsilon_{TW}$.

1.2. Descriptive models

The goal of this section is to offer a simple model of MSM materials. This model has to be able to capture the main phenomena of the material, such as the influence of the magnetic field and mechanical stress on the strain, with a special emphasis on the hysteresis. The connection between magnetic field, stress and strain has been discussed qualitatively in section 1.1.2, where the relevant phenomena where visually illustrated in Figure 1.4 and Figure 1.5. In this section, the influence of the input variables will be determined with more precision, in order to build a simulation model of MSM materials. The model is based on the detailed analysis done in [3], [5], [6] and [7].

1.2.1. Preliminaries

Models of smart materials are generally based on a thermodynamic framework that characterizes the status of the material in terms of its internal energy, and then determines the evolution of the involved variables as an attempt to minimize the total energy. In this section the main concepts for modeling are recalled. The involved variables are, in our case, the magnetic field H , the external stress σ and the strain ε .

The first law of thermodynamics (*energy conservation*) states that, for a unit volume of the material, the change in the internal energy U is given by the sum between the amount $\mathrm{d}W$ of work done by external forces or magnetic fields, and the amount $\mathrm{d}Q$ of heat which enters the material, i.e.,

$$\mathrm{d}U = \mathrm{d}W + \mathrm{d}Q . \tag{1.5}$$

In the case of MSM materials, the contributions to the internal energy are mainly given by the work done by the external stress and the work done by the magnetic field. The work (per unit volume) done by a mechanical stress σ can be quantified as

$$\mathrm{d}W_{MEC} = \sigma \mathrm{d}\varepsilon \ , \tag{1.6}$$

where ε is the deformation or strain. The amount of work (per unit volume) done by a magnetic field on a material is

$$\mathrm{d}W_{MAG} = \mu_0 H \mathrm{d}M \ , \tag{1.7}$$

where μ_0 is the magnetic permeability of the vacuum and M is the magnetization of the material.

The second law of thermodynamics (*irreversible process*) states that the change of the entropy S is

$$\mathrm{d}S = \frac{\mathrm{d}Q}{T} + \mathrm{d}S_{IRR} \ , \tag{1.8}$$

where T is the temperature and $\mathrm{d}S_{IRR}$ represents the entropy produced by *irreversible* processes. Note that $\mathrm{d}S_{IRR} \geq 0$, and when $\mathrm{d}S_{IRR} = 0$ the process is *reversible*. From (1.8) the quantity $\mathrm{d}Q$ can be calculated and substituted into (1.5), that becomes

$$\mathrm{d}U = \mu_0 \mathrm{H} \mathrm{d}M + \sigma \mathrm{d}\varepsilon + T\mathrm{d}S - T\mathrm{d}S_{IRR} \ . \tag{1.9}$$

The internal energy (1.9) is a function of the following independent variables: magnetization M, strain ε and entropy S. For MSM materials another choice of the independent variables is more meaningful, i.e., field H, stress σ and temperature T. The particular choice of the independent variables yields particular formulations of appropriate *thermodynamic potentials*.

A thermodynamic potential is defined as a functional whose minima characterize the equilibrium states of the system in the presence of constraints (or external stimuli). In the context of material characterization, some example of thermodynamic potentials are the Helmholtz free energy and the Gibbs free energy. They constitute thermodynamic potentials for different independent variables and constraints sets.

The Helmholtz potential ψ is defined as

$$\psi = U - TS \ . \tag{1.10}$$

The differential of the Helmholtz potential in case of (1.9) is thus

$$\mathrm{d}\psi = \mu_0 H \mathrm{d}M + \sigma \mathrm{d}\varepsilon - S\mathrm{d}T - T\mathrm{d}S_{IRR} \ , \tag{1.11}$$

which emphasizes that ψ is a function of the independent variables *M*, ε , *T*. It can be noted that for fixed temperature, strain and magnetization regimes,

$$\mathrm{d}\psi = -T\mathrm{d}S_{IRR} \ , \tag{1.12}$$

i.e., the potential ψ is non-increasing ($\mathrm{d}S_{IRR} \geq 0$). The minima of ψ with respect to σ , *H* and *S* give the values of the stress, field and entropy to which the material converges as equilibria. Those minima can be found by

$$\frac{\partial \psi}{\partial H} = 0 \ , \ \frac{\partial \psi}{\partial \sigma} = 0 \ , \ \frac{\partial \psi}{\partial S} = 0 \ , \tag{1.13}$$

which quantify the manner through which the magnetic field, the stress and the entropy adjust so that the internal responses of the material balance the external stimuli. It is noted in [8] that (1.13) yields the necessary conditions

$$H = \frac{1}{\mu_0}\left[\frac{\partial \psi}{\partial M}\right]_{\varepsilon,T}, \quad \sigma = \left[\frac{\partial \psi}{\partial \varepsilon_{EL}}\right]_{M,T}, \quad S = \left[\frac{\partial \psi}{\partial T}\right]_{\varepsilon,M}, \tag{1.14}$$

which offer a way to calculate the resulting field, stress and entropy as variations of the Helmholtz potential due to magnetization, strain and temperature. However, for normal actuation operation, the independent variables of the MSM material should be the field, the stress and the temperature, since they can be chosen independently and constitutes the input to the MSM material system.

To change the set of variables to a more appropriate one, the Gibbs potential G can be defined by

$$G = \psi - \sigma\varepsilon - \mu_0 HM \,, \tag{1.15}$$

which has a differential

$$\mathrm{d}G = -\mu_0 M\mathrm{d}H - \varepsilon\mathrm{d}\sigma - S\mathrm{d}T - T\mathrm{d}S_{IRR} \,, \tag{1.16}$$

where it is pointed out that the independent variable are now H, σ and T. For fixed magnetic field, stress and temperature regimes, (1.16) becomes

$$\mathrm{d}G = -T\mathrm{d}S_{IRR} \,, \tag{1.17}$$

i.e., the potential G is non-increasing ($\mathrm{d}S_{IRR} \geq 0$). The minima of G with respect to M, ε and S give the values of magnetization, elastic strain and entropy to which the material system converges as equilibria. In mathematical terms, the equilibrium equations

$$\frac{\partial G}{\partial M} = 0 \,, \quad \frac{\partial G}{\partial \varepsilon} = 0 \,, \quad \frac{\partial G}{\partial S} = 0 \,, \tag{1.18}$$

quantify the manner through which the magnetization, the elastic strain and the entropy adjust so that the internal responses balance the external stimuli. It is noted in [8] that (1.18) is equivalent to the following conditions

$$M = -\frac{1}{\mu_0}\left[\frac{\partial G}{\partial H}\right]_{\sigma,T}, \quad \varepsilon = -\left[\frac{\partial G}{\partial \sigma}\right]_{H,T}, \quad S = \left[\frac{\partial G}{\partial T}\right]_{\sigma,H}, \tag{1.19}$$

The conditions (1.14) and (1.19) are also called *constitutive relations* of the material.

The thermodynamic potentials and their definitions have been introduced, with particular emphasis to the case of MSM materials. However, no expression of those potentials has been given. In fact, the particular expression depends on the "shape" that is *chosen* for the energies that are involved in (1.10) or (1.15). This is a *choice*, that can be driven by the experience on the particular material, or by some experiments over the system that has to be modeled. For instance, the Helmholtz potential of the spring can be formulated by measuring the liner stress-strain relationship, and then building a ψ in such a way that it has a constant derivative with respect to the strain. For simple systems, this appears a useless procedure. However, for nonlinear systems the presented thermodynamic framework shows its major feature of being quite general and powerful for modeling purposes. Another important fact can be noted: the thermodynamic framework does not require any linearity. It can entail several different effects. It can also deal with irreversible processes. The term $\mathrm{d}S_{IRR}$ has not been addressed in detail here. It will be shown that it plays a fundamental role in the description of MSM materials, through the application of the second law of thermodynamics to quantify the energy losses produced by the hysteresis.

1.2.2. A simple model of MSM materials

This section specializes the thermodynamic framework for the case of MSM materials.

First, let us recall that the overall strain of the MSM element depends on two components,

$$\varepsilon = \varepsilon_{TW} + \varepsilon_{EL}, \tag{1.20}$$

where ε_{EL} is the elastic contribution connected to σ through the Young modulus E of the material, and ε_{TW} is the strain due to variants reorientation. The *twinning strain* ε_{TW} depends on the evolution of the variant V_1. If V_1 is not present in the crystal, then $\varepsilon_{TW} = 0$; if V_1 is the only variant, then $\varepsilon_{TW} = \varepsilon_0$. The quantity of V_1 in the crystal can be expressed by the *variant volume fraction* ξ. When $\xi = 0$ then $\varepsilon_{TW} = 0$, and $\xi = 1$ means $\varepsilon_{TW} = \varepsilon_0$. In other words, the following relationship holds true:

$$\varepsilon_{TW} = \varepsilon_0 \xi. \tag{1.21}$$

Due to the definition of ξ, the volume fraction of the other variant V_2 is $(1-\xi)$. In order to determine the evolution of ε_{TW}, the evolution of ξ must be determined.

The Helmholtz potential of the MSM material is defined by (1.10). Its differential is

$$\mathrm{d}\psi = \mu_0 H \mathrm{d}M + \sigma \mathrm{d}\varepsilon - S\mathrm{d}T - T\mathrm{d}S_{IRR} = \mu_0 H \mathrm{d}M + \sigma \mathrm{d}\varepsilon_{EL} + \sigma \mathrm{d}\varepsilon_{TW} - T\mathrm{d}S_{IRR}, \tag{1.22}$$

assuming that T is constant and using (1.20). Note that the assumption of a constant T is a strong one in the case of MSM material. Temperature plays a fundamental role in the input-output behavior of the material, but complicates the model development. To obtain a simple model, the influence of temperature is here ignored (however, temperature is adequately considered when developing control schemas for MSM-based actuators). A change of independent variables can be done by defining the following Gibbs potential

$$G = \psi - \sigma\varepsilon_{EL} - \mu_0 HM \ , \tag{1.23}$$

which leads to the differential [6]

$$\mathrm{d}G = -\mu_0 M\mathrm{d}H - \varepsilon_{EL}\mathrm{d}\sigma + \sigma\mathrm{d}\varepsilon_{TW} - T\mathrm{d}S_{IRR} \ . \tag{1.24}$$

Equation (1.24) points out that the Gibbs potential is the function of σ, ε_{TW}, H. The irreversibility described by $T\mathrm{d}S_{IRR}$ is taken into account by rewriting (1.24) as follows:

$$-\mathrm{d}G - \mu_0 M\mathrm{d}H - \varepsilon_{EL}\mathrm{d}\sigma + \sigma\mathrm{d}\varepsilon_{TW} \geq 0 \, , \tag{1.25}$$

which constitutes a constraint for the thermodynamic framework to be physically consistent, and is equivalent to (1.17). Note that ε_{TW} is a function of ξ through (1.21), and therefore G in (1.24) is a also a function of ξ.

The time evolution of the potential $G = G(H, \sigma, \xi)$ can be written as a function of the time evolutions of its independent variables, i.e.,

$$\frac{\mathrm{d}G}{\mathrm{d}t} = \frac{\partial G}{\partial H}\frac{\mathrm{d}H}{\mathrm{d}t} + \frac{\partial G}{\partial \sigma}\frac{\mathrm{d}\sigma}{\mathrm{d}t} + \frac{\partial G}{\partial \xi}\frac{\mathrm{d}\xi}{\mathrm{d}t} \, . \tag{1.26}$$

Substituting (1.26) into (1.25) and noting that $\sigma\mathrm{d}\varepsilon_{TW} = \sigma\varepsilon_0\mathrm{d}\xi$, one obtains

$$\left(-\mu_0 M - \frac{\partial G}{\partial H}\right)\dot{H} + \left(-\varepsilon_{EL} - \frac{\partial G}{\partial \sigma}\right)\dot{\sigma} + \left(\sigma\varepsilon_0 - \frac{\partial G}{\partial \xi}\right)\dot{\xi} \geq 0 \, . \tag{1.27}$$

Using the necessary conditions (1.19), equation (1.27) simplifies to

$$\left(\sigma\varepsilon_0 - \frac{\partial G}{\partial \xi}\right)\dot{\xi} \geq 0\,, \tag{1.28}$$

with the following constitutive relationships:

$$\varepsilon_{EL} = -\frac{\partial G}{\partial \sigma}\,, \tag{1.29}$$

$$M = -\frac{1}{\mu_0}\frac{\partial G}{\partial H}\,. \tag{1.30}$$

Equations (1.29) and (1.30) quantify the evolution of the magnetization and the elastic strain which balances the external stimuli given by the field and the stress. In the model that is obtained here, which aims at offering a stress-field-strain connection, the evolution of the magnetization is not relevant, thus (1.30) will be ignored. The quantity $\sigma\varepsilon_0 - \partial G / \partial \xi$ establishes the evolution of the volume fraction, which in turns determines ε_{TW}. For this reason,

$$f^{\xi} = \sigma\varepsilon_0 - \partial G / \partial \xi \tag{1.31}$$

is defined the *thermodynamic driving force* associated to the volume fraction. Note that, even if f^{ξ} is commonly called force, it has the units of a pressure. The constraint (1.28) now reads

$$f^{\xi}\dot{\xi} \geq 0\,, \tag{1.32}$$

which must be satisfied to ensure the consistence of the model with the second law of thermodynamics.

However, to give a sense to (1.28), (1.29) and (1.32), a particular expression of the potential (1.23) has to be chosen. Those energies have to be functions of the independent variables, i.e., σ, ε_{TW} or ξ, and H. A possible choice of G is the following:

$$G = -\frac{1}{2E}\sigma^2 - \mu_0 \int_0^H M(h)\mathrm{d}h + \frac{1}{2} E_{TW}\varepsilon_{TW}^2 , \tag{1.33}$$

where E is the Young modulus of the MSM material, E_{TW} is the Young modulus during variants reorientation (see the slope of the stress-strain characteristic in Figure 1.4 when $\mathrm{V}_2 \to \mathrm{V}_1$ or $\mathrm{V}_1 \to \mathrm{V}_2$). The first term of the potential represents the elastic co-energy, the second the magnetic co-energy of a material in a magnetic field, and the last term the elastic energy due to the interaction between the variants during the reorientation. We remark that the expression of the potential is a choice of the user, and several choices are possible.

Before proceeding, it is necessary to discuss in detail the magnetic term. It is the characteristic one for MSM materials, since the main activation process is magnetically induced. It involves the magnetization-field relationship of the material. Refer to Figure 1.6, which depicts some typical magnetization-field characteristics obtained at different values of ε_{TW}. In particular, the curve $M_1(H)$ is obtained when the material is completely elongated, i.e., $\xi = 1$ since only the variant V_1 is present. The curve $M_2(H)$ is obtained in contraction, such that $\xi = 0$ since only V_2 is present. If the material is in an intermediate strain position, $0 < \xi < 1$, then the typical magnetization curve $M(H)$ develops between the two extreme curves $M_1(H)$ and $M_2(H)$. It can be noted that $M_1(H)$ reaches the saturation value M_S at low magnetic fields, since V_1 has its easy axis parallel to the field and is thus “easy” to magnetize. On the other hand, the curve $M_2(H)$ reaches the saturation value at a much higher value of the field, denoted H_S, since in that case the magnetic field magnetizes the material along its hard axis. We recall that here the magnetic field is assumed to be only along the y direction, and the strain/displacement only along the x direction.

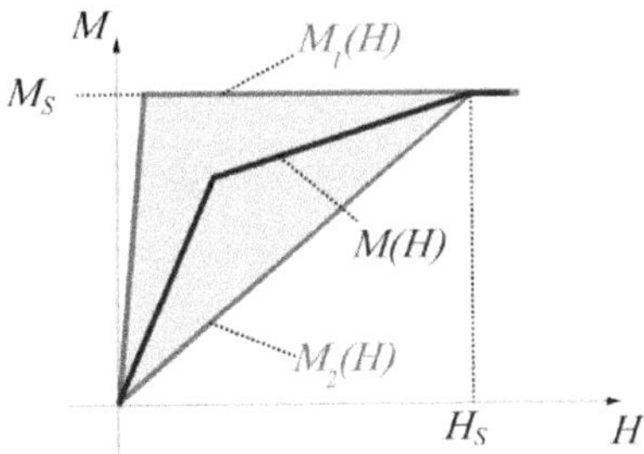

Figure 1.6 – Magnetization versus field strength: curve at elongation $M_1(H)$ (only V_1 is present); curve at contraction $M_2(H)$ (only V_2 is present); generic curve $M(H)$ obtained by fixing the strain of the material in an intermediate position (V_1 and V_2 coexist in the material)

It has been experimentally verified in [3] that an approximate but accurate mathematical expression of $M(H)$ can be the linear (convex) combination of $M_1(H)$ and $M_2(H)$,

$$M(H) = \xi M_1(H) + (1-\xi) M_2(H) . \tag{1.34}$$

In some works $M_1(H)$ is approximated by the saturation value, i.e., $M_1(H) = M_S$ (this corresponds to an infinite magnetic permeability of the easy axis). Equation (1.34) offers the mathematical relationship between field and magnetization that is necessary for (1.33). The magnetic term of (1.33) becomes

$$\mu_0 \int_0^H M(h) \mathrm{d}h = \mu_0 \int_0^H [\xi M_1(h) + (1-\xi) M_2(h)] \mathrm{d}h . \tag{1.35}$$

Now, exploiting the constitutive equation (1.29) and the chosen expression of G in (1.33), the elastic strain can be obtained as a function of the stress,

$$\varepsilon_{EL} = \frac{\sigma}{E}. \tag{1.36}$$

The consequent step is to determine the driving force f^{ξ}, which from (1.31) becomes

$$f^{\xi} = \underbrace{\sigma\varepsilon_0}_{f_{MEC}} + \underbrace{\mu_0 \int_0^H [M_1(h) - M_2(h)]\mathrm{d}h}_{f_{MAG}} - \underbrace{E_{TW}\varepsilon_0^2 \xi}_{f_{MEC}^{TW}}, \tag{1.37}$$

where it is clear that the total driving force is the sum of the mechanical driving force f_{MEC} which depends on σ, the magnetic driving force f_{MAG} which depends on the field *H*, and the mechanical driving force f_{MEC}^{TW} due to the elastic interactions between the variants. It is worth noting that, apart from the scaling factor μ_0, the maximum value of f_{MAG} is graphically represented by the silver area between $M_1(H)$ and $M_2(H)$ in Figure 1.6, and it is obtained when $H \geq H_S$. Actually, this value represents the difference in energy that is required to magnetize the material along the hard axis or the easy axis, i.e., the anisotropy energy. The maximum value of f_{MAG} corresponds to the anisotropy constant K_U of the material.

Using the approximation $M_1(H) = M_S$, the anisotropy constant of the material can be approximated by the expression

$$K_U = \frac{1}{2}\mu_0 H_S M_S, \tag{1.38}$$

and the magnetic driving force can be calculated as ([9])

$$f_{MAG}(H) = \begin{cases} \mu_0 M_S H \left(1 - \frac{\mu_0 M_S H}{4K_U}\right) & H < H_S \\ K_U & H \geq H_S \end{cases} . \quad (1.39)$$

Note that the mechanical driving force is $f_{MEC} = \varepsilon_0 \sigma$ as anticipated in (1.3). However, the total driving force f^{ξ} should satisfy (1.32). So, to complete the model, a relationship between f^{ξ} and ξ has to be *chosen* in such a way that (1.32) holds true at any time. A simple but effective choice is proposed in Figure 1.7 [5]. It represents an analogous situation of what already discussed for Figure 1.5. The change of the volume fraction and thus the variants reorientation starts only when the total driving force overcomes the twinning force f_{TW}. The characteristic of Figure 1.7 is a mean to calculate ξ from the known value of f^{ξ}. To illustrate how this can be done, let us assume that $f^{\xi} = f_A$ at a certain time. Refer to Figure 1.7. Since the characteristic is quasi-static, in the same instant it can be written that $f^{\xi} - f_{TW} = 0$ because of the equilibrium of all the acting forces. The equilibrium condition becomes

$$\begin{aligned} & f_{MEC}(\sigma) + f_{MAG}(H) - E_{TW}\varepsilon_0^2 \xi - f_{TW} = 0 \Rightarrow \\ & \qquad \Rightarrow \xi = \frac{f_{MEC}(\sigma) + f_{MAG}(H) - f_{TW}}{E_{TW}\varepsilon_0^2}, \end{aligned} \quad (1.40)$$

which gives a way to calculate the volume fraction if the increasing path of the characteristic is traversed. The increasing path is characterized by $f^{\xi} \geq f_{TW}$ and consequently $\dot{\xi} \geq 0$.

Let us consider that $f^{\xi} = f_B$ in the decreasing path. The equilibrium condition is now $-f^{\xi} - f_{TW} = 0$, and it leads to

$$
\begin{aligned}
&-f_{MEC}(\sigma)-f_{MAG}(H)+E_{TW}\varepsilon_0^2\xi-f_{TW}=0\Rightarrow \\
&\qquad\Rightarrow \xi=\frac{f_{MEC}(\sigma)+f_{MAG}(H)+f_{TW}}{E_{TW}\varepsilon_0^2},
\end{aligned}
\tag{1.41}
$$

which allows the calculation of the volume fraction along the decreasing path, which has $-f^{\xi}\leq -f_{TW}$, $\dot{\xi}\leq 0$.

The fact that the volume fraction starts increasing or decreasing can be easily checked. If $f^{\xi}-f_{TW}<0$ the volume fraction does not increase, and if $-f^{\xi}-f_{TW}<0$ the volume fraction does not decrease. As soon as $f^{\xi}-f_{TW}>0$ then $\dot{\xi}>0$, or as soon as $-f^{\xi}-f_{TW}>0$ then $\dot{\xi}<0$, and the new value of ξ must be calculated exploiting the equilibrium conditions (1.40) or (1.41), respectively. Note that the conditions are mutually exclusive, in the sense that if $\dot{\xi}>0$ and thus $f^{\xi}-f_{TW}>0\Rightarrow f^{\xi}>f_{TW}$, then $-f^{\xi}-f_{TW}<0\Rightarrow f^{\xi}>-f_{TW}$ must hold.

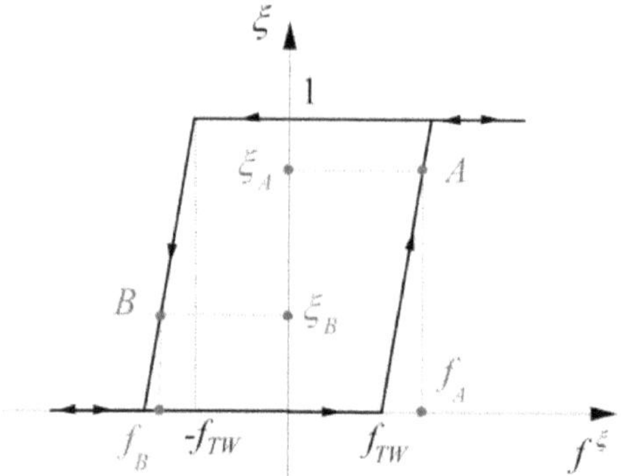

Figure 1.7 – A possible relationship between driving force f^{ξ} and ξ that is consistent with the second law of thermodynamics, satisfying (1.32)

At this point, the model of the MSM material, relating the inputs σ and H to the output ε, is ready. We summarize now how the equations discussed in this section can be used in order to model the MSM material.

1. From the value of σ, the mechanical driving force can be calculated by (1.3).
2. From the value of H the magnetic driving force is obtained by(1.39).
3. The total driving force acting on the twin boundaries is calculated through (1.37), with the present value of ξ.
4. The reorientation conditions are checked; if $f^{\xi} - f_{TW} > 0$ then the variants rearrange $V_2 \rightarrow V_1$ and the new value of ξ is calculated by (1.40). If $-f^{\xi} - f_{TW} > 0$ the variants reorient $V_1 \rightarrow V_2$; the new (lower) value of ξ is calculated by (1.41).
5. The new value of ξ generates a new value of ε_{TW} with (1.21).
6. The elastic strain is obtained by (1.36).
7. The output strain ε is calculated by using (1.20).

Some remarks are necessary. The twinning force f_{TW} is related to the twinning stress by the known proportional expression $f_{TW} = \varepsilon_0 \sigma_{TW}$. The Young modulus E can be different from V_1 to V_2. A common way to deal with this detail is to consider an average Young modulus $E = 0.5E_{V_1} + 0.5E_{V_2}$.

1.2.3. Validation of the simple model

In this section the simple model is verified by generating the field-strain curves at different stress values. Figure 1.8 shows the simulation results.

The twinning stress of the material is $\sigma_{TW} = 0.5$ MPa, which is a realistic value since for common MSM crystal $\sigma_{TW} \in [0.5, 1]$ MPa. The first curve (dashed) is generated for $\sigma = -0.4$ MPa, so it is smaller then σ_{TW}. In fact, it can be seen from Figure 1.8 that the strain is not restored. It remains constant once it reaches the maximum value $\varepsilon_0 = 0.058$.

The second curve (solid) is generated for $\sigma = -0.5$ MPa. In this case, the external stress is able to restore the strain only partially. Note in fact that the total force acting on the boundaries (1.37) has a component related to the elasticity between the variants, f_{MEC}^{TW}, which is maximum when $\xi = 1$ and zero when $\xi = 0$. This contribution depends on the value of E_{TW} and, summed to σ, overcomes the twinning stress producing the little restoring shown by the solid curve of Figure 1.8. The value of E_{TW} affects the slope of each curve of Figure 1.8, as expected from (1.40), (1.41).

The third curve (with triangular marker) corresponds to $\sigma = -1.5$ MPa. The strain is restored. However, the external stress cannot be arbitrarily high. Let us recall that the maximum magnetic stress σ_{MAG} is determined by (1.39) through (1.4). In this example $\max\{\sigma_{MAG}\} = 2.7$ MPa. The curve with the square marker in Figure 1.8 is generated with $\sigma = -2$ MPa, and it can be noted that the maximum strain is not reachable. For $\sigma = -2.5$ MPa, the MSM element does not exhibit any relevant strain. The parameters values used in the simulations are listed in Table 1.1.

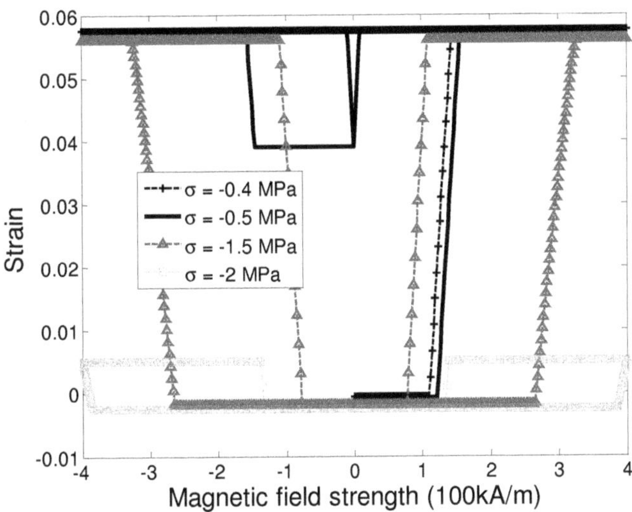

Figure 1.8 - Field-strain curves at different constant stress values; the twinning stress is $\sigma_{TW} = 0.5$ MPa.

Table 1.1 – Parameters of the MSM material used for simulations

Parameter	**Symbol**	**Value**	**Units**
Twinning stress	σ_{TW}	0.6	MPa
Twinning driving force	f_{TW}	$f_{TW} = \sigma_{TW}\varepsilon_0$	MPa
Maximum crystallographic strain	ε_0	0.058	/
Permeability of the vacuum	μ_0	$4\pi \times 10^{-7}$	Tm/A
Anisotropy constant	K_U	1.67×10^5	J/m^3
Average Young modulus	E	800	MPa
Twinning Young modulus	E_{TW}	5	MPa
Saturation magnetization	M_S	5.17×10^5	A/m

1.3. Pro and contra of MSM materials

The working principle of the material has been explained in details, and a quasi-static model has been discussed. In this section, the focus is on some quantities that play a fundamental role in the selection of the appropriate material for a specific application. Generally speaking, these quantities allow a comparison among several active materials, and also several other technologies. The goal of this section is to give to the user an idea of what should be taken into consideration when deciding whether the MSM material is suitable for his application or not.

1.3.1. Meaningful quantities

Actuator technologies can be very different. There are some commonly accepted basic parameters to compare different technologies [10]. They relate to important needs of the applications. In this section, some of those parameters are introduced. A special emphasis is given to those ones which are fundamental for positioning applications.

Certainly, from the application point of view, an important quantity is the *maximum strain* of a material. It is the maximum allowable deformation that the particular material technology can reach. In positioning applications, this is a key-feature.

Another important quantity is the *maximum stress*, which represents the maximum (active) force that can be generated by the active element. Alternatively, it is possible to find the *blocking stress*, which is the minimum stress that does not allow any deformation of the material. The maximum stress and the blocking stress have equivalent values.

The *work output* is an estimation of the maximum work per volume that can be done by the material in a single-stroke. It is calculated as the product between the maximum stress and the maximum strain. A material that has a bigger work output is able to perform the required task in less volume.

The *Young modulus* of the material expresses its stiffness or elasticity, giving an idea of what will be the force response to externally-induced deformations. This modulus defines the passive behavior of the material. The stiffness, the density and the mass of a material determine the resonant frequency of the material system, and have an impact on the resonant frequency of the material-based actuator.

The operating range of a material is limited by the *maximum frequency* at which the material can operate. The value depends on several factors, most of the times concerning more the design of the excitation part than the intrinsic limits of the material.

The *efficiency* is the ratio between the work input, i.e. the amount of energy that is supplied to the material by external excitation, and the work output. Some papers consider as work output the work done by the magnetic force, which neglects the presence of the twinning stress. Some other papers [9] consider as work output the effective work by subtracting the influence of the twinning stress.

The *response time* quantifies the speed of reaction of the material to external stimuli. This is a key-parameter when the goal is to achieve a complete control over the material deformation.

The *resolution* is another important quantity when designing a positioning application. It represents the minimum amount of displacement (smallest step increment) that can be done by the considered material. A good resolution implies that the material can displace in a more "precise" way.

A critical parameter is the *maximum operating temperature.* Since the required temperature is generally a vital requirement of the application, the choice of the best technology cannot ignore this quantity.

1.3.2. Comparison with other active materials

We propose here a brief comparison with other active materials. The comparison is based on the quantities introduced in section 1.3.1 and is reported in Table 1.2.

Table 1.2 – Comparison among common smart (active) materials

	Units	SMA	Magnetostrictors (terfenol-D)	Piezo (layered)	MSMA
Max. strain	%	5-10	0.2	0.2	5 (10)
Max. stress	MPa	100..700	70	20	2.5 (3)
Work output	kJ/m^3	10^4	112	25	150 (300)
Max. frequency	Hz	10	10^7	10^7	10^4
Response time	s	1	10^{-6}	10^{-6}	10^{-3}
Efficiency	%	5	90	90	90
Resolution	m	10^{-5}	10^{-7}	10^{-7}	10^{-7}
Max. temperature	°C	/	200	150	60
Young modulus	GPa	30..90	25..35	45..60	7.7

It is easily noted that the materials which are suited for positioning applications that require a wide displacement range are thermal shape memories (SMA) and MSMAs. The *maximum strain* is quite huge if compared to the strain offered by other material technologies. At present, SMA exhibit a strain up to 10 %, while MSMAs only 5% for crystals that are commercially available. New MSMAs are under development, and some of them show a 10% - 18% strain (in Table 1.2 the value 10% is in brackets since it refers to future generations of crystals).

The *maximum stress* is unfortunately not the best feature of MSMAs. SMAs have a high maximum stress, piezo-ceramics and magnetostrictive materials have a maximum stress that is one order of magnitude bigger than the maximum stress of MSMAs. For applications which require a high stress, then MSMAs seem not to be good candidates.

An advantage of MSM materials is however offered by the *work output*. MSMAs have a higher value than other materials, except for SMAs. For works which require high energy in a small volume, MSMAs and SMAs can be good candidates.

A clear advantage of MSMAs over SMAs is the *maximum frequency*: in fact, they can be driven up to the kHz range. SMAs suffer from the fact that their activation is thermally controlled, i.e., the material needs to be heated and cooled in order to elongate and contract. This puts a severe limitation over the maximum frequency. The consequence is also in the response time. It is really big for SMAs, while quite small for the other compounds and also for MSMAs.

The *efficiency* plays a fundamental role for energy savings. SMAs have a low efficiency since they are thermally activated. Most of the energy that is given in input through heating goes into losses, and only a small part becomes available for mechanical work. A much better situation is for the other materials, which all have high efficiency (90 % and more). MSMAs convert almost all the energy in input in possible work output (of the magnetic force). The work input is given by the energy introduced into the crystal by means of magnetization, so it is expressed by (1.39) as a function of H. The maximum work input is equal to the anisotropy constant K_U. However, the efficiency reported in Table 1.2 does not take into consideration the losses which are due to hysteresis. Only a part of (1.39) goes actually on the external load, since there is the friction produced by the twinning stress. The hysteresis losses are particularly relevant when a cyclic operation is required from the actuator. The efficiency of MSMAs in cyclic operation has been estimated to be about 50-70 %, and is anyway a function of the particular value of twinning stress of the crystal [11]. A reduction of σ_{TW} will improve the efficiency in cyclic operation.

The *resolution* is the smallest increment in strain that can be obtained from a material, and is very important when dealing with positioning applications [12]. MSMAs have a low resolution, comparable with the ones of piezo ceramics and magnetostrictive materials.

The *maximum temperature* is quite low for MSMAs, and at present this is the main factor which discourages the industrial adoption of such materials in important fields such as automotive. This issue is under extensive analysis from the materials community, which is preparing MSM alloys of the next generation.

From this brief comparison, some conclusions can be made. MSMAs appear as the best candidates for positioning systems, since they have a huge strain, an acceptable active stress to move a small load, a good resolution, a small response time and a good maximum frequency. However, it would be better if the positioning application does not require a cyclic operation regime but single strokes. The MSM-based actuator has to be designed in such a way that the MSM crystal does not experience relevant temperature variations. Moreover: the presence of the twinning stress, that is a disadvantage in cyclic operation, can become a *huge advantage* for positioning applications with single strokes. It allows holding a position without supplying energy, and increases the efficiency of this material with respect to other technologies. This issue is addressed in more details in the next chapter, where MSMA-based actuators are described.

Several research groups are working on improving the production process of MSM materials [13], [14], [15], [16] or their actuation characteristics such as the martensite-to-austenite transition temperature [17]. Thus, we believe that several parameters of MSMAs shown by Table 1.2 will have a better value in the next years.

1.4. Other effects

MSM materials belong to the class of the so-called *smart materials*. They are defined as materials which are "active", in the sense that they are able to act on the external world by doing mechanical work and, at the same time, they are able to react to external stimuli by changing some of their properties. We have extensively discussed how MSM materials are able to act; it is worth mentioning how they react to external excitations.

MSM materials react to the application of an external stress by contracting. However, contraction is not the only phenomenon. Refer to Figure 1.6. It shows the behavior of the magnetization versus field. This behavior is completely determined by the value of the volume fraction, i.e., the field-magnetization curve depends on the actual value of twinning strain. The field-magnetization curve expresses the magnetic permeability μ_{MSM} of the MSM element, which is a function of the twinning strain, $\mu_{MSM} = \mu_{MSM}(\varepsilon_{TW})$. This relationship can be used to sense the strain by measuring μ_{MSM} [18], [19]. This inherent sensory effect, which is a particular feature of all smart materials [20], can be exploited within *self-sensing* strategies, where one active material is used to actuate and to sense at the same time.

The change in permeability can be used in another way. If the material is excited by a fixed magnetic flux density *B*, then the change in permeability produces a variation of the inductance *L* of the material. This produces an induced voltage proportional to the term $\mathrm{d}L/\mathrm{d}t$. In other words, an external load which applies cycling forces to the MSM element produces a cycling voltage, and thus produces *energy*. There is an active research area which studies the possibilities of the use of MSM material within actuators whose goal is *energy harvesting* [21]. Recently, the Finnish company Adaptamat released a new commercial actuator with this specific goal. The big energy density of the material and the big magnetic anisotropy make MSM alloys good candidates for energy harvesting applications.

Another effect of MSM materials relies on the fact that they are electrically conductive. A change in their strain corresponds to a change in their electrical resistivity. However, there are two drawbacks to the exploitation of this effect: to measure the resistivity, an electrical circuit must be connected to the element, and it complicates the design of actuators; moreover, the resistivity is obviously influenced by temperature. Note that also the change in permeability is affected by temperature. The main difference is that the permeability is less sensitive than resistivity, since the field-magnetization curve does not change significantly if the temperature variations are bounded.

2. MSM actuators

A MSM actuator is an electro-mechanical device which is able to excite the MSM element to obtain elongation and/or contraction, and thus the displacement of a load. This chapter introduces some design concepts of MSM actuators. It has to be mentioned that, due to the relatively recent discovery of MSM materials, the actuator design represents a very active research area. Questions such as "what is the best actuator structure for a specified application" still have no answers.

In the past years, research about MSM materials has grown. Several actuator designs have been proposed: some of them are able to exploit features that are unique for MSM materials. In this chapter some concepts will be described in a concise way.

Section 2.1 introduces the *operating modes* of MSM actuators: they are basic schemes of how a MSM material can be used within an actuator in order to produce reversible motion. Section 2.2 presents two actuator structures. The first is nowadays the most commonly used in the MSM actuator community and takes the name of "spring" actuator. A prototype made by the Finnish company Adptamat and belonging to this category is also commercially available for research or applications. The second structure, called "push-push" or *multistable* actuator, is able to exploit some features of the MSM material in an intelligent way, and in our opinion is extremely interesting for new positioning systems. Several other design are possible, and the interested reader is referred to [22] and [23] for a broader overview.

Section 2.3 discusses some issues regarding MSM actuators which are relevant from the control viewpoint. In particular, a possible modeling is proposed. This modeling aims at being effective for control design, so it has been chosen to be as simple as possible: the actuator is represented as the series between a hysteresis operator, which represents the MSM element, and a linear/nonlinear dynamic, which represents the mechanical load that the element has to move. Another relevant issue concerns the influence of temperature on the hysteresis. By the experimental data acquired from two different prototypes, it is shown that temperature affects the hysteretic characteristic significantly. Section 2.3 thus emphasizes the main points that a control algorithm for positioning purposes must take into consideration: the presence of hysteresis, and the fact that this nonlinearity is time-varying.

2.1. Operating modes

With *operating mode* we intend a particular way to deal with the MSM material within an actuator. The modes are based upon different way to exploit the physical effect of the material as explained in section 1.1. Any MSM actuator belongs to one category defined by an operating mode. The operating modes discussed next have been summarized for the first time in [24].

The operating modes are shown in Figure 2.1. The operating mode 1 refers to those kinds of actuators which achieve contraction by means of mechanical stress, and elongation by applying a magnetic field through the MSM element. The mechanical stress is generally produced by a spring that loads the active element. The stiffness of the spring has to be chosen carefully, since the elastic force should be able to contract the element despite the twinning stress. At the same time it should not be too big, since otherwise the magnetic field would not be able to elongate the element. When the magnetic field increases along the *y* direction, the generated magnetic stress along *x* pushes the twin boundary to favor V_1. When the magnetic field is decreased, the spring pushes the boundary back. At present, this operating mode is the most common for MSM actuators (see section 2.2.1).

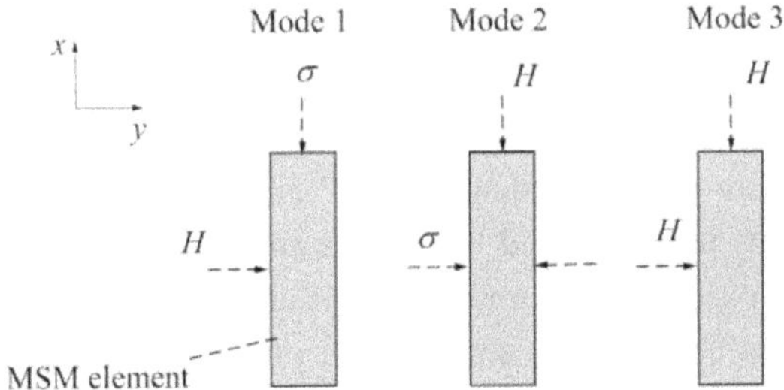

Figure 2.1 – Operating modes of MSM actuators

The operating mode 2 refers to those actuators where elongation is achieved by mechanical stress along the *y* direction, whereas contraction is driven by the applied magnetic field along the x direction. Actually, this actuator concept is very rare in literature. An example can be found in [25], a master thesis done in collaboration with LPA.

The third operating mode is maybe the most interesting one for MSM actuators. In this case, elongation and contraction are driven by the application of magnetic fields (e.g., perpendicular to each other). There are several advantages of this operating mode, which are discussed next.

When thinking to a specific operating mode, the reader should keep in mind that the generation of a magnetic field along a direction requires the presence of coils (or permanent magnets). This complicates the mechanical design of the actuator. For example: the operating mode 1 requires the presence of coils along the y direction, which means that the actuator has a big volume in the surroundings of the element. On the other hand, mode 2 requires a coil along the x direction, and this coil can be placed around the MSM element itself. This reduces the volume of the coil. However, the fact that a mechanical stress has to be applied along y complicates the design of the coils along x.

The operating mode 3 looks really attractive. At first sight, it also requires a complicated design, since two coils have to be placed along y and x respectively. A particularly innovative design is proposed in [22] and [26]. The efforts in the design phase are definitely paid back by the advantages. First of all, the deformation in both directions is controllable. The active force of the MSM element (see section 1.2.2 and equation (1.39)) can be used to "push" the load for elongation, and to "pull" the load when contracting. In mode 1, for example, the active force can push, but the contraction is driven by the spring. That is, the maximum force in contraction corresponds to the maximum value of the elastic force of the spring. In mode 2, the mechanical stress is (usually) also generated by some elastic elements, and the issue repeats itself just on a different direction. A second advantage concerns the amount of active force which is transferred to the load. In mode 1, the magnetic force has to overcome the sum of the twinning stress and the external stress generated by the pre-stress spring, which changes with the displacement. In mode 3, the active force has to deal only with the twinning stress: all the rest goes to the mechanical load that has to be moved (which can also entail a spring-like behavior). Another important advantage of mode 3 is the exploitation of the twinning stress. If the field is switched off, the deformation tends to remain at the same value, since the twinning stress opposes to any motion. This means that mode 3 can offer an actuator design that is efficient from the energetic viewpoint, and robust with respect to the action of disturbance forces. A particular hybrid case of mode 3 is the "push-push" actuator: elongation and contraction are driven magnetically because two MSM elements are arranged antagonistically, one being the load for the other. This concept is detailed in section 2.2.2.

In the next section two types of actuators are detailed, the "spring" actuator and the "push-push" actuator, respectively from operating mode 1 and operating mode 3.

2.2. Some actuator structures

2.2.1. The MSM "spring" actuator

The most common type of MSM actuator is the "spring actuator". It belongs to the category of actuators which work in operating mode 1. The mechanical stress that contracts the material is generated by a spring. The stress exerted by the spring is bigger than the twinning stress. A schema of the spring actuator is depicted in Figure 2.2.

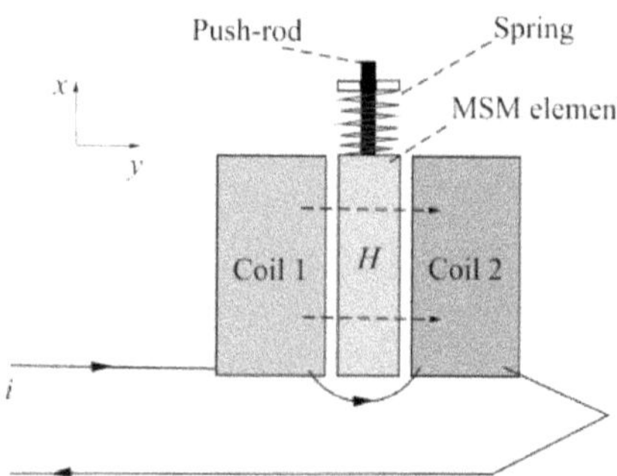

Figure 2.2 – The MSM spring actuator

The magnetic field which elongates the material is generated by two coils connected in series and placed aside of the active element. They are traversed by a current i produced by a current source (such as a voltage-controlled current amplifier). The deformation of the MSMA is transmitted to the push-rod, which represents the mechanical interface with the external load that has to be moved.

Figure 2.3 shows the picture of a prototype of spring actuator. The prototype has been developed by LPA at Saarland University in Saabruecken, Germany. It is possible to clearly distinguish the two excitation coils; the MSM element is in the middle and is not visible. The push-rod is not attached to any spring. The spring has to be added externally. Another spring actuator is shown in Figure 2.4, which is the picture of the commercial version produced by Adaptamat (the coils are inside the housing). The spring is collocated inside the protection of the push-rod. The spring and the coils are not visible. The manufacturing is particularly optimized to disfavor the creation of eddy currents while driving the coils. Another prototype, developed by TU Braunschweig, is shown by Figure 2.5. The spring is visible and integrated in the design.

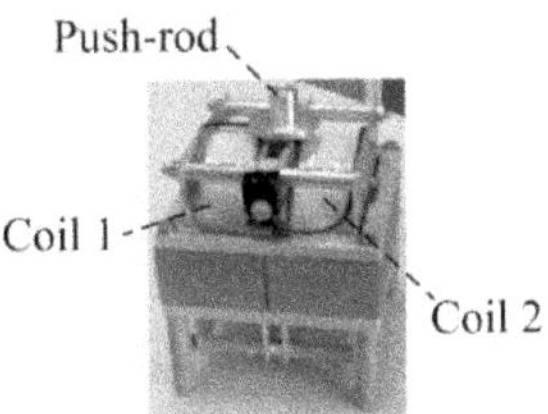

Figure 2.3 – Prototype of spring actuator (from LPA, Saarland University)

Figure 2.4 – Prototype of spring actuator from Adaptamat (Helsinki, Finland)

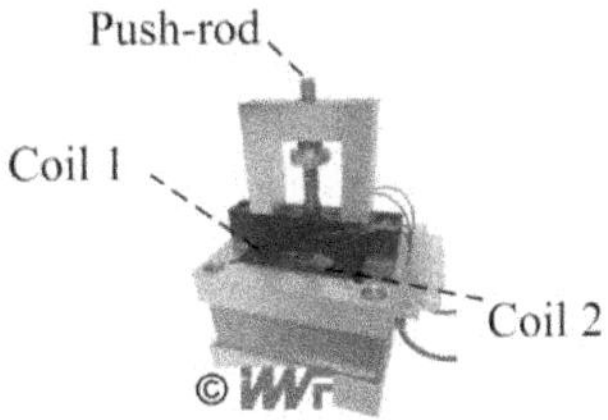

Figure 2.5 – Prototype of spring actuator from TU Braunschweig (Braunschweig, Germany)

2.2.2. The "push-push" actuator

This interesting actuator concept was proposed in [27]. It is composed by two MSM elements (A and B) arranged antagonistically, as shown in the schematics of Figure 2.6.

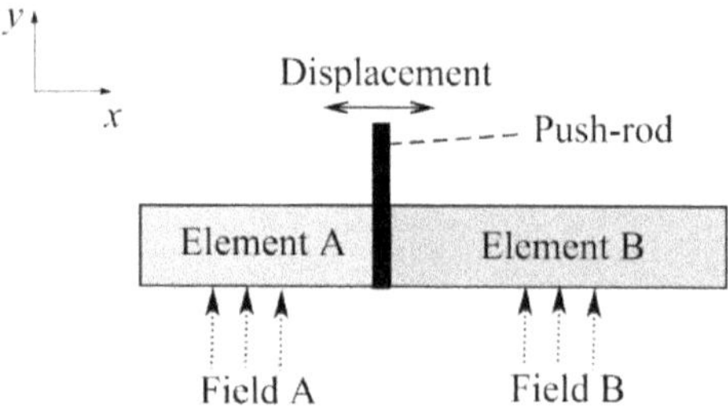

Figure 2.6 – Schematic view of a MSM push-push actuator

The working principle is as follows. A magnetic field is applied through the element A in order to obtain positive displacements along x, contracting element B; a magnetic field is applied through element B in order to displace in the other versus. The magnetic field of each element is generated by independent excitation coils. The push-push actuator can be seen as a particular case of operating mode 3, since contraction and elongation are controlled magnetically. It is worth noting that the push-push is also able to exploit the twinning stress as a feature: once the push-rod is moved, its position can be held thanks to the twinning stress of both active elements. An external force (e.g. a disturbance force) must overcome the two twinning stresses of the two elements in order to start a movement. To emphasize this feature, the push-push is also called *multistable actuator*, since any intermediate position is an equilibrium position of the device. A prototype of push-push is presented in Figure 2.7 (developed by ETO MAGNETIC GmbH, Stockach, Germany). The interested reader can find in [28] a modeling approach for this actuator, based on the thermodynamical framework of section 1.2, and in [29] a control approach for positioning purposes.

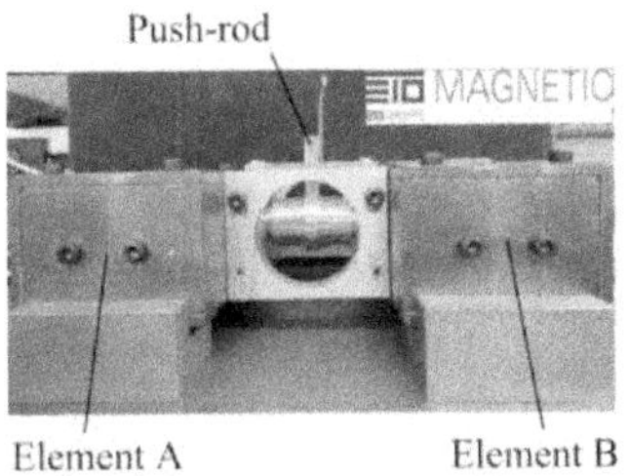

Figure 2.7 – Prototype of MSM push-push actuator (by ETO MAGNETIC GmbH)

2.3. Control issues

The main goal of this thesis is to discuss position control by means of MSM-based actuators. In section 2.2 some possible designs of MSM actuators for positioning have been illustrated. In this section the focus is on how a MSM actuator can be considered from the viewpoint of a closed-loop, where it is the component which produces motion. The discussion made here is kept quite general to comprehend any kind of MSM actuator listed before.

2.3.1. Modeling for control

A MSM actuator is composed by three main parts: the magnetic circuit, the MSM element and the mechanical load, which is connected to the active element via some interface and is the part that has to be moved for positioning purposes. A schema of these components and their mutual interactions is reported in Figure 2.8.

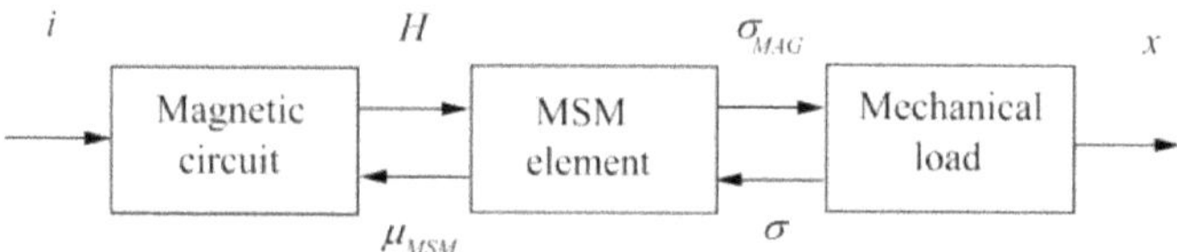

Figure 2.8 – Interactions between the components of a MSM actuator

The magnetic circuit is composed mainly by the excitation coils that, when traversed by the current i, generate the magnetomotive force within the circuit. The magnetic field H through the MSM element is consequently generated as a function of the current i and of the magnetic reluctances of the circuit. The relationship between i and H can be assumed to be linear, through a coefficient that depends on the involved magnetic reluctances [9].

The input current i is usually generated by a voltage-controlled current source. The current source does not belong strictly to the MSM actuator, so it is not shown in

Figure 2.8. However, the reader should consider the importance of this component: it is a closed-loop system which controls the current output i to make it follow the desired current profile in input. The quality of the control depends on the dynamics of the loop, which are introduced by the elements of the magnetic circuit. For instance, the dimension and other properties of the excitation coils modify the overall inductance which is seen by the current source. When the internal controller is tuned well, the desired current and the output current i coincide in the frequency band of interest. In this case, the presence of the current source can be neglected for the model.

The output of the magnetic circuit is the field H through the MSM element. The element itself can be represented as the generator of the magnetic active stress σ_{MAG} (see (1.39)), which acts on the mechanical load. In turn, the mechanical load exerts a stress σ on the MSM element, opposing or favoring the deformation. The mechanical load is the "object" that the actuator has to move. It can entail some elastic effects, for instance in the case where a loading spring is present (as for the spring actuator of section 2.2.1). The interaction between the load and the MSM element determines the actual displacement of the load, x, which is the quantity to be controlled.

If the load is connected to the MSM element, the displacement x is related to the strain ε of the material through

$$x = \varepsilon l_0 , \tag{2.1}$$

where l_0 is the length of the MSM material in complete contraction. We recall that ε is made by the elastic strain plus the twinning strain. The latter influences the magnetic permeability μ_{MSM} of the material which, in turn, influences the relationship between the current i and the field H. The variation can be neglected if a good current source is designed in order to have a low sensitivity to those permeability changes which occur during actuation. Note that since $\varepsilon - H$ is hysteretic, also $x - H$ will be so.

The precise modeling of the interactions depicted in Figure 2.8 is an extremely difficult task. The paper [30] proposes a really detailed model of a MSM actuator working in operating mode 1. The model is based on the Hamiltonian formalism, which can make use straightforwardly of the thermodynamic framework presented in section 1.2.1. The paper [31] presents another model that is simpler in the description of the magnetic circuit. If, on one hand, both modeling approaches showed to be really effective for the correct dimensioning of a MSM actuator and for the simulation of its dynamic behavior, they have three drawbacks when referring to the application. The first concerns the precision of the resulting models. They rely on an approximate description of the hysteresis between field and twinning strain, which is similar to the one shown by Figure 1.7, and which is not able to describe the *complex* hysteresis that arises in the material (refer to section 3.1.2 for a clarification). The second drawback concerns the usability of the models for control purposes. Since they present several state variables, nonlinear and hysteretic relationships, the choice of a control strategy is not trivial. The third drawback is that both approaches need knowledge of the actuator system and of its parameters, e.g. the Young modulus, anisotropy constant and twinning stress of the MSM material. Those parameters are not easy to measure and they vary from one MSM element to another. From the user point of view, it could be advantageous to have a modeling approach which is based on input-output experimental analysis and which is directly usable for control design, in the sense that this activity can rely on established results, or the development of a new one is simplified.

A common representation of smart actuators is presented in Figure 2.9 [32], [33], [34]. It is quite general and can embrace several kinds of actuators. In the case of MSM actuators, the input u is the current i through the excitation coils. The hysteresis Γ is due to the presence of the MSM material; the output v can be seen as a hysteretic force generated by the active element. This force excites the mechanical load that is described by linear dynamics (e.g. the mass-spring-damper dynamics). The resulting displacement x depends on the balance between the load and the active force coming from the MSM element.

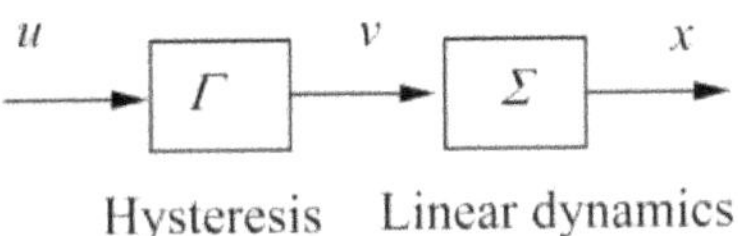

Figure 2.9 – Representation of smart actuators

The description of Figure 2.9 contains some approximations in the case of MSM actuators. First of all, the influence of x on the input u, which happens through the change in permeability, is ignored. The approximations offer a clear way to think about MSM actuators from the control viewpoint, and has been successfully used in [35]. A control algorithm should take into consideration two main phenomena of the actuator: the hysteresis and the dynamics. In the general case, the latter can also be considered as nonlinear.

2.3.2. Influence of temperature

When temperature increases, the MSM material returns in the austenite phase where any (magnetically-induced) actuation is impossible. In other words, temperature reduces the strain of the material. To illustrate the phenomenon, some experimental data are presented. Figure 2.10 shows how the increase of temperature reduces the hysteretic input-output characteristic of the actuator developed at Saarland University (Figure 2.3). At room temperature T_0 the actuator offers a displacement peak-to-peak range of about 180 μm. At $T_1 > T_0$, the displacement range reduces to about 80 μm. A further increase of temperature up to $T_2 > T_1$ brings the available motion within 30 μm.

Another example is shown by Figure 2.11 and concerns the actuator from Adaptamat (Figure 2.4). In this case, it is possible to appreciate a particular property of MSM materials. Refer to the curve acquired at $T_1 > T_0$. The MSM element is able to offer more strain than at room temperature (note that the difference is almost 200 μm!). This is due to the fact that the twinning stress reduces with increasing temperature. However, a further increase then reduces the available strain because of the growth of the austenite volume fraction, despite of the martensitic one.

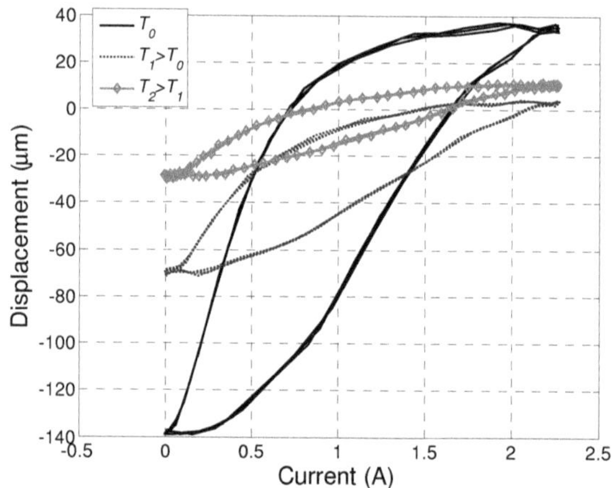

Figure 2.10 – Influence of temperature on the actuator of Figure 2.3

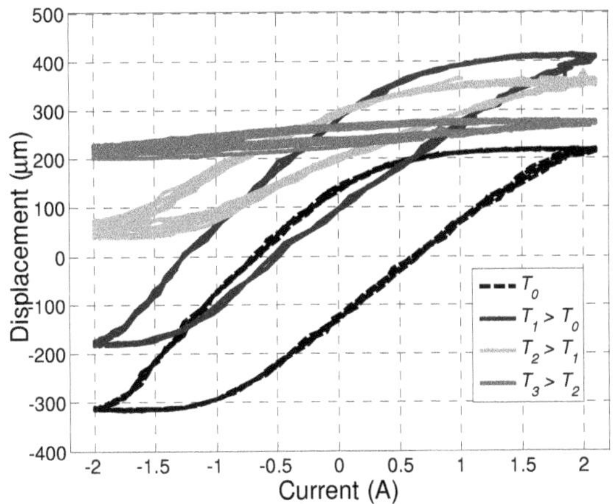

Figure 2.11 – Influence of temperature on the actuator of Figure 2.4

This points two things out. The first is that there exists an optimal temperature at which MSM actuators can work, offering the maximum displacement. The second is that that the controller must take into consideration a time-varying hysteresis. A more precise representation of MSM actuators is depicted in Figure 2.12.

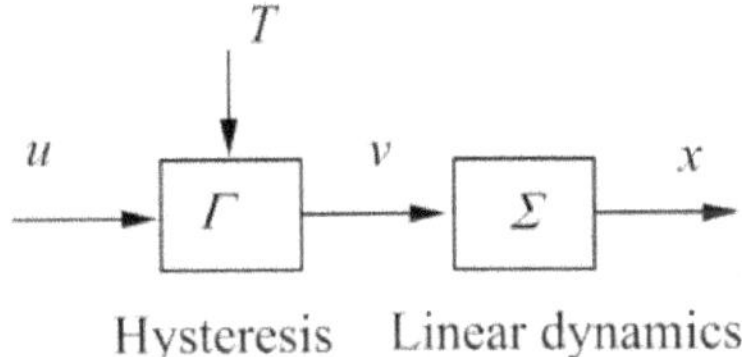

Figure 2.12 – Representation of a MSM actuator with emphasis on the time-varying behavior of its hysteresis

The representation of Figure 2.12 will be taken as the reference for the development of the control algorithms.

3. Hysteresis

Hysteresis nonlinearities are nonlinear behaviors which arise in a wide variety of processes where the input-output dynamic relations involve some memory effects. Examples of hysteresis are found in biology, magnetism and mechanical structures (bridges, building…), where hysteresis is a natural mechanism of materials to supply forces against movements, and to dissipate energy. Hysteresis is, in fact, often related to *dissipation phenomena* inside a structure or a material.

Several smart materials and smart-material-based systems exhibit the hysteresis behavior between the input variable, such as the electric field for piezo-ceramics or the magnetic field strength for magnetostrictive materials, and the output variable, especially when they are driven with sufficiently high amplitudes. Thus, a common approach to face the presence of hysteresis, is to reduce the operating range of the system, in such a way that it operates in a quasi-linear regime (where the nonlinearity can be neglected).

However, this approach is not meaningful for MSM systems. In this case, the outstanding strain is the characterizing property of the MSM material, and reducing the operating range does not exploit its main potentiality. Since the hysteretic phenomenon cannot be neglected, it is necessary to develop methods that face the presence of the hysteresis and, at the same time, are amenable to control design.

The detailed modeling of hysteretic systems by means of physical laws is quite challenging, or the obtained models are then too complex to be used in practice. For this reason, several modeling approaches have been proposed. Some of them do not rely on a detailed physical knowledge of the phenomenon, but rather attempt to offer a mathematical tool that combines some basic physical understanding of the hysteretic system along with a kind of black-box modeling.

They are called *semi-physical* or *phenomenological* models. Those models generally offer a way to describe hysteresis, independently of the particular context (mechanical, biological, chemical…) where the hysteresis arises.

Some examples of phenomenological models are the well-known *Bouc-Wen, Preisach*, *Prandtl-Ishlinskii*, and *Krasnosel'skii-Prokrovskii* models. The Bouc-Wen model has been used with success for modeling mechanical structures. These elements exhibit an inelastic behavior, producing a restoring force that depends on the complete history of their deformation, and not only its instantaneous value. The Bouc-Wen model is able to capture the memory nature of such a restoring force. The features of the Bouc-Wen model are its dependence on a few parameters, and the fact that it consists of "just" one nonlinear differential equation. On the other hand, the drawbacks concern the difficulty in the identification of the parameters, the nonexistence (at present) of an inverse model, the simplicity of the described hysteresis (for example, it is symmetric with respect to the zero point). The recent monograph [36] reports mathematical results which address the first issue; some papers, e.g. [37] face the last one; but the impossibility to invert the model is still an open problem for some applications.

In fact, in some cases it is necessary to adopt a hysteresis model that is amenable for control design, and especially for the *compensation* (or *inversion*) of the hysteretic nonlinearity. To better understand the need of compensation, the reader is referred to the Wiener or Hammerstein models of systems, which include the presence of a *static nonlinearity* in the input or output path of a feedback-loop. A common way to control those systems relies on the adoption of an inverse model of the nonlinearity.

Among the cited hysteresis models, the *Preisach*, *Prandtl-Ishlinskii*, and *Krasnosel'skii-Prokrovskii* models (called *Preisach-like* models since they share a common framework) are tools which come along with the corresponding inverses or *compensators*. The inversion of such models is straightforward thanks to the huge amount of literature devoted to them. Moreover, the particular mathematical formulation allows adopting simple procedures for the identification of the parameters of those models, and at the same time offers a way to model curves which are asymmetric or (almost) arbitrarily complex.

This chapter focuses on the Preisach-like models. Section 3.1 introduces the hysteresis phenomenon from a mathematical perspective. Then introduces the two main classes of hysteresis, defined with respect to their memory structure: the *local* (*noncomplex*) and *nonlocal* (*complex*) hysteresis nonlinearities. Section 3.2 illustrates the classical Preisach hysteresis model and then focuses on the Krasnosel'skii-Pokrovskii model. The main issues concerning the modeling and the inversion of the hysteresis nonlinearity are discussed. Section 3.3 focuses on the modified Prandtl-Ishnlinskii model: the issues of model formulation, identification and inversion are addressed. This model is the main tool that is used in chapters 4 and 5, devoted to the position control problem. Section 3.4 presents some properties of the hysteresis models, in particular the Lipschitz continuity and the Lipschitz constant of the Krasnosel'skii-Pokrovskii and the modified Prandtl-Ishlinskii models.

3.1. What is hysteresis?

3.1.1. Introduction

Hysteresis is a nonlinear behavior with memory of an output variable with respect to an input variable. Let us consider a hysteretic mapping between an input u and an output v, as shown by Figure 3.1. Let us assume that, at the beginning of the experiment, the input has value $u_0 = u(t_0)$. The output value $v_0 = v(t_0)$ must belong to the admissible set $\Sigma = \Sigma(u_0)$ (depicted as a grey vertical line in Figure 3.1). Let the point $\mathrm{E} = (u_0, v_0 = 0)$ describe the initial point on the hysteretic curve. If the input is now increased from u_0 to u_1, the output follows the path until it reaches F. If then the input continues increasing up to u_2, the output still follows the path towards the point G. Another possibility is that the input is decreased from u_1 to u_3: in this case, from the point F the output moves towards H. The *branching* of the u-v map is basically due to the memory effect entailed by a hysteretic nonlinearity. It is a consequence of the fact that the value of the output $v(t)$ depends on the complete history of $u(.)$ and not only on $u(t)$, as it happens for *static* nonlinearities. Obviously, each point in the u-v plane must belong to the *hysteretic region* Ω, which is the area enclosed by the major loop.

Another typical characteristic of hysteresis is that it does not depend on the rate of the input signal. As consequence, the mapping shown in Figure 3.1 will persist for any frequency of the input u. This is called *rate-independent property.* In [38] hysteresis is curiously defined as "the ghostly image of the input-output map as the frequency of the excitation goes to zero", to emphasize that hysteresis is neither a dynamic phenomenon neither a static one.

The branching and looping of a ferromagnetic hysteresis was investigated by Madelung at the beginning of the 20th century. He determined some properties which can qualitatively describe the phenomenon (refer to Figure 3.2):

1. Any curve C_1 emanating from a turning point A of the input-output trajectory is uniquely determined by the coordinates of A;
2. If any point B on the curve C_1 becomes a new turning point, then the curve C_2 originating at B leads back to A;
3. If C_2 is continued beyond A, then it coincides with the continuation of the curve C which led to the point A before $C_1 - C_2$ was traversed.

In addition to those properties which belong to almost any hysteresis characteristic, there is a fourth important one playing an important role for the hysteresis of smart-materials.

4. (*crossing-loop property*). More than one curve can traverse a non-turning point D. The branch that led to D is uniquely determined by the relevant past history of the input signal *u*.

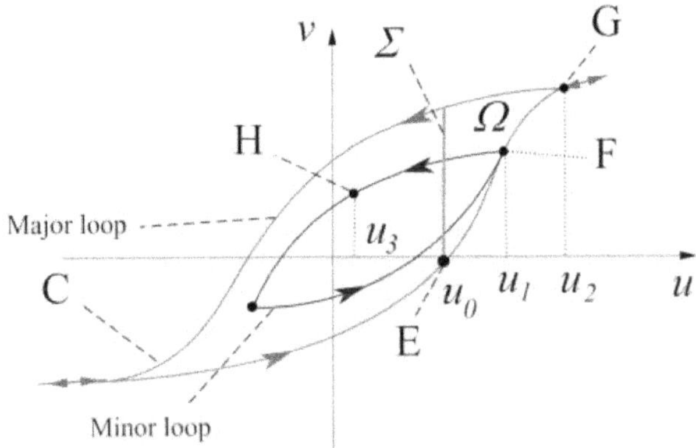

Figure 3.1 – An example of hysteretic mapping with major and minor loops

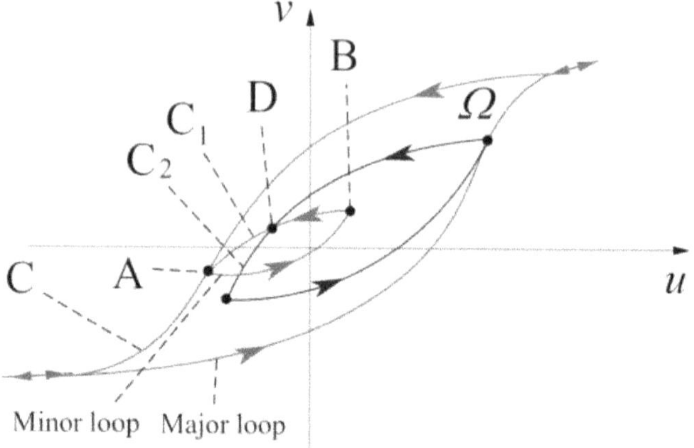

Figure 3.2 – Madelung's rules for hysteresis

3.1.2. Noncomplex and complex hysteresis

Hysteresis phenomena can be *noncomplex* (*local*) or *complex* (*nonlocal*). The definition mainly refers to the particular memory structure of the phenomenon.

Hysteresis nonlinearities with local memory are particular hysteresis for which the pair $(u_0, v_0) \in \Omega$ and the values of $u(t)$ for $t > t_0$ completely (and uniquely) determine $v(t)$. In other words, the influence of the input on the output, up to t, is considered through the momentary value $(u(t), v(t)) \in \Omega$. The state of a hysteresis with local memory is scalar, and can be represented by just a point on the u-v plane. A noncomplex hysteresis $\Gamma[\cdot]$ can be expressed as

$$v(t) = \Gamma[u; u_0, v_0](t), \tag{3.1}$$

where $u(\cdot), u_0 \in \Re$, $v_0 \in \Sigma(u_0)$ and $t \geq t_0$.

Hysteretic nonlinearities with nonlocal memory are characterized by the fact that the pair $(u_0, v_0) \in \Omega$ and the values $u(t)$ for $t > t_0$ are not enough to determine the future evolution of $v(t)$. In this case, also the values of $u(t)$ for $t < t_0$ influence $v(t)$ in $t > t_0$. This influence is considered through the definition of an internal state of the hysteresis, which can be also of infinite dimension. If the state is of finite dimension, it can be captured by a state vector. Thus, a complex hysteresis $\Gamma[\cdot]$ can be expressed as

$$v(t) = \Gamma[u; z_0](t) , \qquad (3.2)$$

where the vector $z_0 \in \Re^n$, with possibly $n \to +\infty$, is the internal state.

To provide an example, let us consider again the hysteresis depicted in Figure 3.2, assuming to start at t_0 in the point D. A further change of the input u for $t > t_0$ can let the output move on the curve C_1 or on the curve C_2. It is evident that the initial information given only by the pair (u_0, v_0), coordinates of D, is not sufficient to determine which minor loop will be traversed with $u(t)$. Thus, the hysteresis in Figure 3.2 is a complex/nonlocal hysteresis. In fact, if a hysteresis has the crossing-loop property, then it is complex.

3.1.3. Mathematical description

In mathematical literature, hysteresis is defined as a *causal rate-independent* operator that maps continuous functions of time to continuous functions of time. In the following those properties are established in a more formal way.

Let $\Re_+$ be the set of non-negative real numbers.

Definition 3.1. *A function* $f \in C(\Re_+)$ *is ultimately non-decreasing (increasing) if there exists a* $\tau \in \Re_+$ *such that f is non-decreasing (increasing) in* $[\tau, +\infty)$.

Definition 3.2. *A function* $f : \Re_+ \rightarrow \Re_+$ *is a time transformation if f is continuous, non-decreasing and* $\lim_{t \to +\infty} f(t) = +\infty$.

Definition 3.3. *An operator* $\Gamma : C(\Re_+) \rightarrow C(\Re_+)$ *is a causal operator if, for all* $\tau \geq 0$ *and all inputs* $u_1, u_2 \in C(\Re_+)$, *the fact that* $u_1 = u_2$ *in* $[0, \tau]$ *implies that* $\Gamma[u_1] = \Gamma[u_2]$ *in* $[0, \tau]$. *Equivalently, such an operator is said to have the Volterra property.*

Definition 3.4. *An operator* $\Gamma : C(\Re_+) \rightarrow C(\Re_+)$ *is rate-independent if, for every time transformation f such that* $f(0) = 0$ *and* $f(T) = T$, *the following holds:*

$$\Gamma[u \circ f](t) = \Gamma[u](f(t)), \ \forall t \in [0, T]. \tag{3.3}$$

Definition 3.5. *A causal, rate independent operator* $\Gamma : C(\Re_+) \rightarrow C(\Re_+)$ *is a hysteresis operator.*

In this brief subsection the main mathematical properties of a hysteresis operator have been formally defined. The next section starts listing the precise mathematical formulation of some hysteresis operators which are widely used in practice for modeling and control design purposes.

3.2. Preisach-like models

In subsection 3.1 the hysteresis phenomenon has been introduced. No precise mathematical characterization or expression of the operator Γ has been given. This is due to the fact that the mathematical expression depends on the particular choice of the hysteresis model. In this section we focus on a subset of a specific model family called Preisach-like models.

The Preisach model has its genesis in the context of ferromagnetism in the 1935. Preisach proposed a concept for modeling the hysteresis between field strength and magnetization as the superposition of elementary kernels, which were called *hysterons*. In the 1970's, Krasnosel'skii and his collaborators pointed out that the concept encapsulated a rich mathematical framework, amenable to be studied from an abstract perspective. His studies emphasized that the Preisach model was a powerful mathematical tool, able to describe hysteresis in a wide range of scientific contexts. One strength of the Preisach concept is the fact that is based on hysterons. Those entities are elementary hysteretic kernels that can be defined in a variety of ways, depending on the specific necessity. One drawback of the Preisach concept is that it cannot entail straightforwardly the physics of the phenomenon. Present work is focused on this issue [8].

In section 3.2.1 the classical Preisach operator is introduced. Section 3.2.2 deals with a particular Preisach-like model called Krasnosel'skii-Pokrovskii operator, which can describe also the reversible effects of a hysteresis. The identification and the inversion issues are discussed.

3.2.1. The classical Preisach operator

The classical Preisach model (PM) is based on the hysteron shown in Figure 3.3, also called relay operator, which is parameterized by two thresholds values α and

β, $\alpha \leq \beta$. This operator describes a noncomplex hysteresis through the mathematical expression

$$v(t) = k[u; v_0](t) = \begin{cases} k[u; v_0](0) & \text{if } \tau(t) = 0 \\ -1 & \text{if } \tau(t) \neq 0 \text{ and } u(\max(\tau(t)) = \alpha \\ +1 & \text{if } \tau(t) \neq 0 \text{ and } u(\max(\tau(t)) = \beta \end{cases}, \tag{3.4}$$

where the initial value $k[u; v_0](0)$ is defined as

$$k[u; v_0](0) = \begin{cases} -1 & \text{if } u(0) \leq \alpha \\ v_0 & \text{if } \alpha < u(0) < \beta \\ +1 & \text{if } u(0) > \beta \end{cases}, \tag{3.5}$$

and the set $\tau(t)$ takes into consideration the instants at which the input u crosses the thresholds,

$$\tau(t) = \{\bar{t} \in [0, t] \mid u(\bar{t}) = \alpha \text{ or } u(\bar{t}) = \beta\}. \tag{3.6}$$

Note that (3.4) and (3.5) depend on an initial state v_0, which coincides with the output value just before the input $u(0)$ is provided. The output value itself captures all the memory since the relay operator has local memory as specified in section 3.1.2. The output $v(t)$ is then also the internal state of the hysteron: it is completely reset once a switch occurs, as emphasized by (3.4). The evolution of the internal state/output is determined completely by the last threshold or switch.

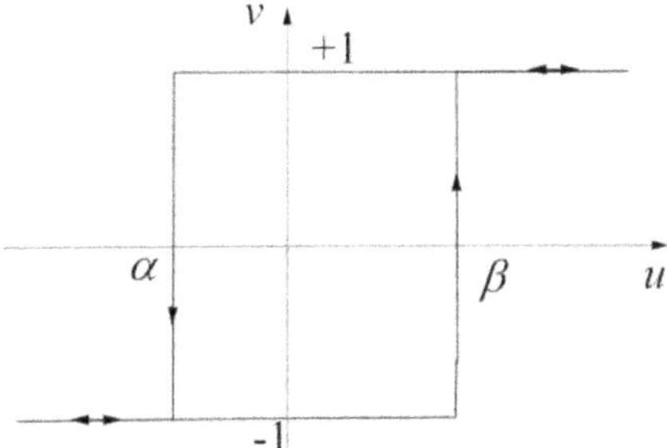

Figure 3.3 - Classical Preisach hysteron

The relay operator describes a simple, elementary hysteresis. A macroscopic, complex hysteretic behavior $\Gamma[\cdot]$ such as the one in Figure 3.1, is represented by the superposition of many hysterons $k_{\alpha,\beta}[\cdot]$ characterized by different thresholds values, as follows

$$v(t) = \Gamma[u; z_0](t) = \iint_P w(\alpha,\beta)\, k_{\alpha,\beta}\left[u; z_0(\alpha,\beta)\right] \mathrm{d}\alpha \mathrm{d}\beta \,, \tag{3.7}$$

where $w(\alpha,\beta)$ denotes a density or *weight function* on the Preisach (half) plane P which is

$$P = \{\alpha, \beta \in \Re \mid \alpha \le \beta\} \,, \tag{3.8}$$

and is represented in Figure 3.4.

There is a one to one correspondence of the hysterons with points in the plane P. Each hysteron can be graphically represented as a point in P. The initial state is now captured by $z_0 = z_0(\alpha,\beta)$, which is a memory curve with support on P. The state of each hysteron is given by the value that the memory curve assumes in the corresponding point of P.

The weight function determines the "strength" of the contribution of a particular hysteron $k_{\alpha,\beta}[\cdot]$ to the overall output through the weight $w(\alpha,\beta)$ in a point of *P*. In section 3.1.1 the concept of hysteretic region has been introduced. This region does not need to be closed. However, in most of the cases which are relevant in practice, Ω is a compact set enclosed by the major loop. On the Preisach plane, this is described by a weight function $w(\alpha,\beta)$ that has compact support $\bar{P} \subset P$, where

$$\bar{P} = \{\alpha, \beta \in \Re \mid \alpha_{MIN} \leq \alpha \leq \beta \leq \beta_{MAX}\}, \tag{3.9}$$

and is represented by the silver area in Figure 3.4.

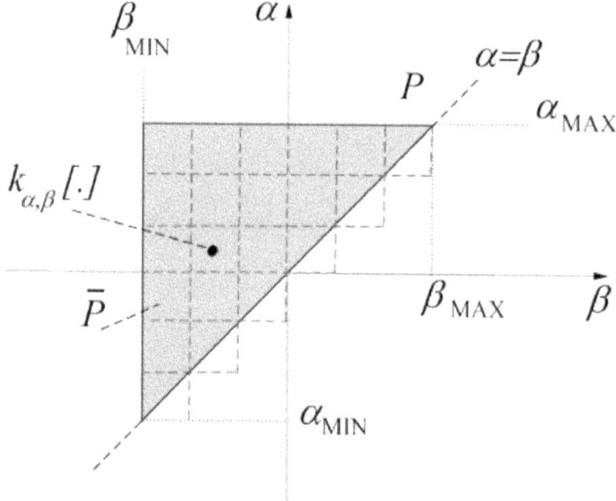

Figure 3.4 – The Preisach plane *P* and the hysterons as points in *P*

The model in (3.7) is not amenable for computation since it involves a continuum of hysterons, i.e. an infinite number of them. The model can be discretized by means of a discretization of level *L* of the Preisach plane. This means that the value range of the thresholds α and β is divided in *L* parts, leading to a number

$K = L(L+1)/2$ of hysteron after the discretization. The grid in Figure 3.4 shows an example of discretization of order 6. The weight function $w(\alpha,\beta)$ is consequently approximated by the expansion w_K,

$$w_K = \sum_{i=1}^{K} w_i \delta_i \, , \tag{3.10}$$

where *i* indicates a node in the discretized plane (corresponding to one hysteron), w_i is the weight of the node and δ_i is the Dirac measure having center at the *i-th* node. Roughly speaking, (3.10) relates a specific hysteron *i* to a specific weight w_i. The discrete Preisach model can thus be expressed as the sum

$$\Gamma[u; z_0] = \sum_{i=1}^{K} w_i k_i [u; z_{0i}](t) \, , \tag{3.11}$$

where $z_0 = [z_{01}, \ldots, z_{0K}]^T$ contains all the initial states, and $k_i[\cdot]$ refer to a precise hysteron characterized by the thresholds pair (α_i, β_i). Note that the discrete operator (3.11) and the original one in (3.7) share the same symbol $\Gamma[\cdot]$. This should not cause confusion, because in the following we will refer only to the discrete version of any hysteresis model.

The classical Preisach operator has some strengths and limitations. It is possible to prove the following theorem, which invokes two important properties: the *minor loop closure* and the *congruency* [39].

Theorem 3.1 [8]. *For piecewise monotone inputs* $u(t)$, *the minor loop closure and the congruency properties constitute necessary and sufficient conditions for a hysteresis operator to be represented by a classical Preisach operator (3.7).*

The minor loop closure and the congruency are then two conditions that could be checked experimentally before describing the phenomenon via a Preisach model.

On the other hand, these conditions are not always met by smart-materials and smart-material-based devices, due to other phenomena (e.g. *relaxation*) that are not captured by the classical Preisach formulation. In those cases, the modeling has to take into consideration other approaches or extended Preisach methods [8].

One limitation that is evident from the kernel definition (3.4) and the vertical "arms" of the hysteron in Figure 3.3 is that it is not able to describe *reversible effects*, such as elastic effects (which would be represented by arms having a slope). This limitation can be overcome with alternative definitions of the kernel.

3.2.2. The Krasnosel'skii-Pokrovskii operator (KPM)

The Krasnosel'skii-Pokrovskii model (KPM) is built as the weighted superposition of the so-called Krasnosel'skii-Pokrovskii hysteron, depicted in Figure 3.5.

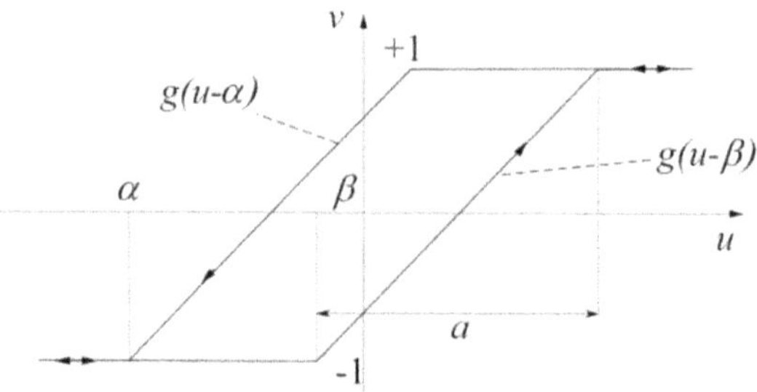

Figure 3.5- The Krasnosel'skii-Pokrovskii hysteron

The superposition which constitutes the KPM is already in (3.7); obviously, the kernel $k_{\alpha,\beta}[\cdot]$ is now the one in Figure 3.5. To express it mathematically, let us consider a piecewise monotone input u that is monotone in each $t \in [t_{j-1}, t_j]$, $j = 1 \ldots J$. Moreover, for brevity let us refer directly to the discretized Preisach plane, where each hysteron is a i-th node in the plane. The i-th KP hysteron is determined by

$$k_i[u;z_{0i}](t) = R[u,R_{j-1}](t) \,, \tag{3.12}$$

where

$$R[u,R_{j-1}](t)\begin{cases} \max\left\{R_{j-1}, g\left(u(t)-\beta_i\right)\right\} & \text{if } u(t) \text{ is non-decreasing} \\ \max\left\{R_{j-1}, g\left(u(t)-\alpha_i\right)\right\} & \text{if } u(t) \text{ is non-increasing} \end{cases}, \tag{3.13}$$

and where the state R_j (at time t_j) is calculated as follows:

$$R_j = \begin{cases} R[u,R_{j-1}](t_j), & j=2,...J \\ R_0 = z_{0i}, & j=1 \end{cases} . \tag{3.14}$$

In (3.13) the *ridge function* $g(\cdot)$ appears, which generally is chosen as

$$g(\chi) = \begin{cases} -1 \text{ if } \chi < -1 \\ 1 \text{ if } \chi > 1 \\ -1+2\dfrac{\chi}{a} \text{ if } -1 \leq \chi \leq 1 \end{cases} . \tag{3.15}$$

Note that the state R_j is changed only at times t_j, $j=1...J$ where the monotone input *u* reaches its minima or maxima over the monotonicity interval $[t_{j-1},t_j]$. The elementary KP operator describes a noncomplex hysteresis. The overall discrete KPM, which models complex hysteresis, is built as in (3.11) by using the new definition (3.12).

Each *i*-th hysteron of the KPM is parameterized by the thresholds pair (α_i,β_i) and the slope-modulation *a*. Generally, *a* has the same value for all of the *K* hysterons composing the KPM.

3.2.3. Identification of the KPM

Once a PM or a KPM structure is chosen, the main issue is how to fit the model to experimental data to describe the hysteresis. From (3.11), it should be clear that the identification aims at calculating the following parameters: the K pairs α_i, β_i, the slope-modulation a and the weights w_i, $i = 1...K$.

Let us assume that the input belongs to the compact set $u(t) \in [\underline{u}, \overline{u}]$ for each $t \in \Re_+$, as happens in practice due to the presence of saturation. The discretization of level L of the Preisach plane can be done using the upper and lower value of the input range,

$$\begin{aligned} \alpha_i &= \underline{u} + (i-1)\Delta u \\ \beta_i &= \underline{u} + (j-1)\Delta u \end{aligned}, \quad i, j = 1...L \text{ and } j \geq i, \tag{3.16}$$

where $\Delta u = (\overline{u} - \underline{u}) / (L-1)$ divides the input range into L intervals. Moreover, it is common to assume that each hysteron share the same constant $a = \Delta u$.

The identification of the weights is more complicated, since they determine the precise shape that the modeled hysteresis will have and thus the quality of the model matching. The thresholds and a impose the structure of the PM or KPM approximators; the weights determine how the hysteron interacts and contribute for the overall output behavior. The weights strictly depend on the input-output experimental data.

Let us consider a set of T input-output data, $u(1)...u(T)$ and $v(1)...v(T)$. The most commonly used method is the linear least squares approach which minimizes the following cost functional

$$F(w) = \frac{1}{2}\sum_{n=1}^{T}\left(v(n) - \sum_{i=1}^{K} w_i k_i[u; z_{0i}](n)\right)^2, \tag{3.17}$$

where the functional F is a function of the weights $w = [w_1, ..., w_K]^T$.

Equation (3.17) points out that another element is necessary for the identification: the knowledge of the initial state vector z_0. Since the input signal can be determined *a priori*, a common approach is to make it starting from the minimum value or the maximum value $\bar{u}$, such that the state of all hysterons is, respectively, $z_{0i} = -1$ or $z_{0i} = +1$ for $i = 1...K$.

Equation (3.17) does not contain any constraints on the weights. This depends however on the goal of the identified model. Generally, the weights are constrained to be positive ($w_i > 0$ for $i = 1...K$), since this ensures the existence of a continuous inverse hysteresis operator for the PM and KPM.

3.2.4. Inversion of the KPM

In many practical applications, an inverse hysteresis model is required to compensate for the effects of hysteresis, which generally produce poor tracking performance. The inversion problem can be stated in the following terms: given a desired output $v_D(t)$ and a hysteretic system $\Gamma[\cdot]$, the inverse model $\Gamma^{-1}[\cdot]$ has to calculate $u^*(t)$ such that $\Gamma[u^*; z_0](t) \simeq v_D(t)$. The situation is sketched in Figure 3.6.

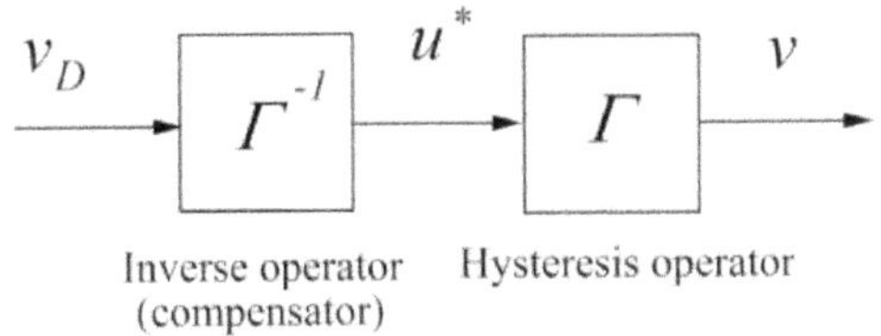

Figure 3.6 - Schema of hysteresis inversion (compensation)

To invert a PM or a KPM, it is first necessary to know several facts. First of all, we should be able to determine if and when a PM or a KPM are invertible. Then, it is of interest to know if the inverse model is continuous. Continuity ensures that small changes in the desired signal v_D correspond to small changes of the inverse output u^*. Finally, the last question concerns how the inverse model looks like from a mathematical viewpoint. The first two issues are addressed by Theorem 3.2.

Theorem 3.2 [39]. *If the weight function $w(\alpha,\beta)$ is positive on a arbitrarily thin strip along the line $\alpha=\beta$ (refer to Figure 3.4), then the KPM and the PM are invertible with a continuous inverse operator.*

Unfortunately, the last question has a brutal answer [39]: the inverse operator of a Preisach operator is not a Preisach operator. Thus, despite the certainty about the existence and continuity, no mathematical structure is given for the compensator. The inversion is commonly done numerically, and the resulting compensator is then a numerical approximation of the unknown one.

A detailed discussion about the inversion can be found in [40]. Here the main result of the algorithm is briefly reported. The theorems that follow are necessary for the inverse schema to work.

Theorem 3.3 [39]. *If the weight function is non-negative, then the Preisach operator $\Gamma[\cdot]$ is piecewise monotone, in the sense that for all monotone functions $u\in C[0,T]$ and for all memory curves z_0,*

$$\left(\Gamma[u;z_0](T)-\Gamma[u;z_0](0)\right)\left(u(T)-u(0)\right)\geq 0\,. \tag{3.18}$$

Theorem 3.4 [39]. *If the weight function is bounded on the Preisach plane, then $\Gamma[\cdot]$ is Lipschitz continuous with constant C_L, such that for all $v_1,v_2\in C[0,T]$ and all memory curves z_0,*

$$\left\|\Gamma[u_1;z_0]-\Gamma[u_2;z_0]\right\|_\infty\leq C_L\left\|u_1-u_2\right\|_\infty\,. \tag{3.19}$$

Note that the hypotheses of both theorems are always met in practical applications. Theorem 3.3 is valid because the weight function is forced to be non-negative (see section 3.2.3) in the identification procedure. Theorem 3.4 applies because a sufficient condition for the weight function to be bounded is that the number of operators is finite, which is always the case for the discrete model (3.11).

For a Preisach operator that is piecewise monotone and Lipschitz continuous, the (approximated) recursive inversion algorithm is as follows

$$\begin{cases} u^{[h+1]}(n+1) = u^{[h]}(n+1) + \dfrac{v_D(n+1) - \Gamma[u^{[h]}(n+1); z_0(n)]}{C_L} \\ u^{[0]}(n+1) = u^{\infty}(n) \end{cases}, \tag{3.20}$$

where n denotes a time-instant, $n = 0...T$, and h denotes a variable used for recursion, $h = 0...N_E$. The value of N_E, which denotes the number of iterations to be done, should be high enough to ensure convergence of the calculated input to the final value $u^{\infty}(n+1)$ that has to be given in output at time n+1. The variable $u^{\infty}(n)$ denotes the input given to the hysteretic system at the previous instant. It can be proved that $\Gamma[u^{[h]}(n+1); z_0] \to v_D(n+1)$ uniformly as $h \to +\infty$, i.e., the inversion algorithm works provided that the number of iterations is sufficiently high.

3.3. The modified Prandtl-Ishlinskii operator

The modified Prandtl-Ishlinskii hysteresis operator (MPIM) is based on the so-called Prandtl-Ishlinskii model (PIM), which is a special version of the Preisach operator in (3.7) [41]. A key-feature of the MPIM is that is admits *an explicit inverse model*, differently from the PM and the KPM, for which an approximated inverse algorithm with recursion was needed (see (3.20)). This is a fundamental property of a hysteresis model. In practical applications the compensator has to be implemented in real-time on the available hardware: the fact that its structure is completely known in terms of computational burden can remove some unnecessary intermediate steps of the hardware design and testing phase.

From a mathematical viewpoint, the MPIM is the composition in series of a PIM and a piecewise-linear function called the Superposition Operator (SO). In the following both components are detailed. For the sake of brevity and to avoid useless repetitions, only the discrete version of the model and its components is discussed.

3.3.1. The Prandtl-Ishlinskii hysteresis model (PIM)

The basic idea of the PIM is to model a complex hysteresis nonlinearity as the weighted superposition of elementary operators called *play* or *backlash* hysterons, depicted in Figure 3.7. In this section, the input to the PIM is denoted with u and the output with υ, while the output of the *j*-th hysteron is denoted with υ_j.

It can be noted that this operator is very similar to the KP operator. However, the reader should note that the KP operator has a saturation effect, i.e., the output value belongs to the compact set $[-1,1]$. The play operator does not have a saturation. The hysteretic region is thus an open set, $\Omega \subset \Re^2$. Another difference is that the j-th play operator is usually characterized by just one threshold parameter r_j.

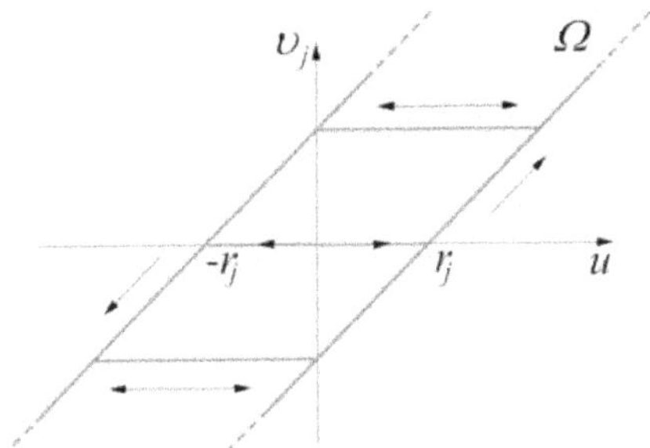

Figure 3.7 - Play operator

Let us consider a piecewise monotone input u with monotonicity partition $t_0 \leq t_1 \leq .. \leq t_{k-1} \leq t_k \leq t_{k+1} \leq .. \leq t_K$. The j-th play operator is mathematically described by the recursive equation

$$\upsilon_j(t) = H_j[u; z_{0j}](t) = R[u; R_{k-1}](t) \text{ , } t_{k-1} \leq t < t_k \tag{3.21}$$

where this time

$$R[u, R_{k-1}](t) = \max\{u(t) - r_j, \min\{u(t) + r, R_{k-1}\}\} \text{ ,} \tag{3.22}$$

$$R_k = \begin{cases} R[u, R_{k-1}](t_k) & k = 2...K \\ R_0 = z_{0j} & k = 1 \end{cases} . \tag{3.23}$$

It can be noted from (3.23) that the state R is updated only at the instants t_k where the input signal changes direction, i.e., at the extremes of the monotonicity intervals of the partition. The hysteretic characteristic of Figure 3.7 is completely determined once the threshold r_j is fixed and an initial state z_{0j} is known.

The play operator describes, as usual for any hysteron, a noncomplex hysteresis. A complex hysteresis such as the one in Figure 3.2 is modeled by the weighted superposition of such hysterons, which define the *discrete* PIM $H[\cdot]$ as

$$\upsilon(t) = H[u; p, r, z_0](t) = \sum_{j=0}^{N} p_j H_j[u; r_j, z_{0j}](t) , \tag{3.24}$$

where z_0 is the vector containing the states z_{0j} of each operator, $r = [r_0, ..., r_N]^T$ is the vector of thresholds, $p = [p_0, ..., p_N]^T$ the vector of weights. The thresholds are arranged in such a way that they compose an ordered sequence $r_0 = 0 \leq r_2 \leq ... \leq r_N$. In matrix notation, (3.24) becomes

$$\upsilon(t) = H[u; p, r, z_0](t) = p^T \underline{H}[u; r_j, z_{0j}](t) \ , \tag{3.25}$$

where $\underline{H}[u; r_j, z_{0j}] = [H_0[u; r_0, z_{00}], ..., H_N[u; r_N, z_{0N}]]$ is a vector containing the output of each elementary operator.

Note that the PIM has been denoted by the operator $H[\cdot]$, differently from the KPM and the PM where the symbol $\Gamma[\cdot]$ has been used. This is due to the fact that the PIM is here intended as one component of a more general model, the MPIM. The MPIM will be denoted by $\Gamma[\cdot]$ in order to have a uniform notation of the hysteresis operators considered in this thesis.

Section 3.3.2 discusses how a PIM can be identified from input-output data.

3.3.2. Identification of the PIM

Let us refer to (3.25): the identification of a PIM is based on the calculation of its parameters, i.e. the weight vector p and the threshold vector r, once an initial state z_0 and the signals $u(t)$ and $v(t)$ are given. We assume that the input-output signals have been acquired for $t \in [t_0, t_E]$.

As for the KPM and the PM, the thresholds are generally fixed *a priori.* A common choice is to place the thresholds with a uniform distribution over the input range $[\underline{u}, \overline{u}]$, splitting it into *N* intervals, where *N* is called the *order* of the PIM,

$$r_j = \frac{j}{N+1} \max_{t_0 \le t \le t_E} \{|u(t)|\}, \quad j = 0 \ldots N. \tag{3.26}$$

The initial states of the hysteron can also be fixed *a priori*, if and only if an appropriate initialization procedure is done. This procedure consists in driving the system with a monotone input in order to bring it in the (0,0) point of the hysteretic characteristic. In this case, it is possible to assume $z_0 = 0$.

Once that states and thresholds have been determined, it can be observed that the PIM (3.25) has a linear-in-the-parameters (LIP) structure with respect to the weights p. However, the reader should recall that the main goal of the identification is to give a model that is amenable for control purposes, and thus for inversion. Consequently, it is of interest to determine which conditions ensure the invertibility of the hysteresis model and thus the existence of an inverse operator, since these constraints must be taken into consideration when carrying out the identification of the model. It turns out that the identification problem is a *constrained* linear problem.

To discuss in more details the invertibility issue, it is necessary to introduce the concept of the *generator function* of the PIM. A method to characterize univocally the hysteresis of a PIM is to look at the so-called *initial loading curve*, which is traversed when the initial state vector z_0 is the zero vector, and the input u is monotonically increasing up to the maximum $\overline{u}$. This curve, denoted by $\varphi(\overline{r})$, can be equated with a threshold-dependent piecewise linear function

$$\varphi(\overline{r}) = \sum_{j=0}^{i} p_j\left(\overline{r} - r_j\right), \quad r_i < \overline{r} < r_{i+1}, \quad i = 0 \ldots N, \tag{3.27}$$

where $r_{N+1} = +\infty$. The curve given by (3.27) is called the generator function of the PIM. Since it characterizes univocally the PIM, an idea to obtain an inverse model is to find that PIM which has a generator function $\varphi_I(\overline{r}_I)$ that is the inverse of $\varphi(\overline{r})$, such that $\varphi_I(\varphi(\overline{r})) = \overline{r}$. The thresholds of the inverse PIM (IPIM) are denoted by r_I. For the inverse function $\varphi_I(\overline{r}_I)$ to exist, the function $\varphi(\overline{r})$ needs to be invertible, and thus strictly monotone. The strict monotonicity holds if

$$\frac{\mathrm{d}\varphi(\overline{r})}{\mathrm{d}\overline{r}} > 0 \tag{3.28}$$

is satisfied. Since $\varphi(\overline{r})$ is piecewise linear, the inequality (3.28) becomes

$$\sum_{j=0}^{i} p_j > 0, \; i = 0 \ldots N. \tag{3.29}$$

The inequality (3.29) ensures that the PIM generated by the weights p_j is invertible. Note that the invertibility depends only on the weights. In this sense, a different choice of the thresholds (3.26) will not affect this fundamental property. If the goal of the PIM identification is to create an inverse model for compensation, then the identification has to consider (3.29).

The identification of the weights of the PIM can then be formulated as the constrained L_2^2-norm minimization of the following prediction error:

$$e_P(p) = p^T \underline{H}[u; z_0] - v, \tag{3.30}$$

which compares the predicted output of the hysteresis $H[u; p, r, z_0](t)$ (see(3.25) with the measured output of the "real" hysteresis $v(t)$. Thanks to the LIP structure of e_P, the optimization is a linear quadratic optimization where the functional to be minimized is

$$F = \min_{p \in \Re^{N+1}} \left\{ \left(\int_{t_0}^{t_E} \left(p^T \underline{H}[u; z_0](t) - v(t) \right) dt \right)^2 \right\}, \qquad (3.31)$$

subjected to the constraints (3.29).

3.3.3. Inversion of the PIM

The inverse PIM, or IPIM, is also called compensator. Its existence is connected to the fact that the PIM is invertible, condition ensured by (3.29). Moreover, we have mentioned that the PIM is completely characterized by its generator function $\varphi(\overline{r})$. The IPIM can be seen as that PIM which is characterized by a generator function $\varphi_I(\overline{r}_I)$ that is the inverse of $\varphi(\overline{r})$, such that $\varphi_I(\varphi(\overline{r})) = \overline{r}$. Moreover, $\varphi_I(\overline{r}_I)$ is completely determined by the thresholds r_I and weights p_I, as in (3.27) after obvious substitutions. Here the formulas to calculate the inverse parameters are reported.

The value of the thresholds can be calculated exploiting the relationship $r_{Ij} = \varphi(r_j)$, obtaining

$$r_{Ii} = \sum_{j=0}^{i} p_j \left(r_i - r_j \right), \; i = 0 \ldots N. \qquad (3.32)$$

The value of the weights can be calculated using

$$p_{I0}=\frac{1}{p_0},\ p_{Ii}=-\frac{p_i}{\left(p_0+\sum_{j=1}^{i}p_j\right)\left(p_0+\sum_{j=1}^{i-1}p_j\right)},i=1..N \quad (3.33)$$

There is a relationship also between the initial states of the direct model, z_0, and of the inverse model, z_{I0}. Under the assumption that $z_0=0$, then also $z_{I0}=0$. This assumption is made all along the manuscript.

Note that (3.32) and (3.33) offer a way to completely calculate an inverse model starting from its direct counterpart. This means that the compensator of a PIM, or a IPIM, can be found *analytically*. In other words, the model generated by (3.32) and (3.33) constitutes an inverse hysteresis operator $H^{-1}[\upsilon;p_I,r_I,z_{I0}]$ such that

$$H\left[H^{-1}[\upsilon;p_I,r_I,z_{I0}],p,r,z_0\right](t)=\upsilon(t), \quad (3.34)$$

for each $\upsilon\in C[0,+\infty)$ and each $t\in\Re^+$.

3.3.4. The Superposition Operator (SO)

The SO is built as the weighted superposition of *2L+1* elementary memoryless operators called half-sided dead-zone operators (DZOs). The *l*-th DZO is depicted in Figure 3.8, where υ is the input, v_l represents the output of the *l*-th DZO, and a_l is the threshold.

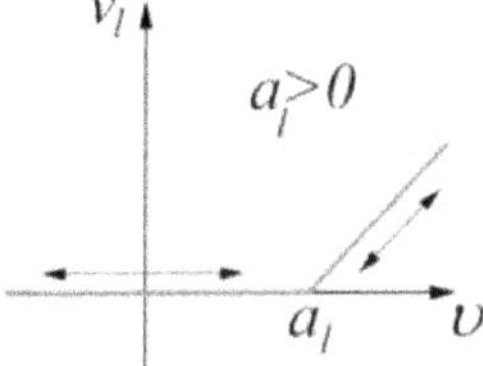

Figure 3.8 – Half-sided dead-zone operator (DZO)

The DZO with $l=0$ has the threshold $a_0=0$, so it corresponds to a linear function, as detailed by (3.35). The mathematical description of the l-th DZO is

$$v_l(t)=S_l\left(v(t);a_l\right)=\begin{cases}\max\{v(t)-a_l,0\} & a_l>0\\ \min\{v(t)-a_l,0\} & a_l<0\,.\\ v(t) & a_l=0\end{cases} \tag{3.35}$$

The overall SO $S(\cdot)$ is composed by the weighted superposition of DZOs, i.e.,

$$S\left(v(t)\right)=\sum_{l=-L}^{L}w_l S_l\left(v(t);a_l\right)=w^T\underline{S}\left(v(t);a\right), \tag{3.36}$$

where $a=\left[a_{-L},\ldots,0,\ldots,a_L\right]^T$ is the vector of thresholds and $w=\left[w_{-L},\ldots,w_0,\ldots,w_L\right]^T$ is the vector of the weights.

Note that the SO is also rate-independent because of the particular definition of the DZO in (3.35). Moreover, it presents several important structural analogies with the PIM, which are really useful for the identification and the inversion.

3.3.5. Inversion of the SO

The inversion is not discussed in details. However, it relies on the same considerations already done in section 3.3.2 and 3.3.3. First of all, the SO has to be invertible, and this comes from similar constraints on the weights, i.e.,

$$\sum_{l=-L}^{0} w_l > 0, \quad \sum_{l=0}^{L} w_l > 0. \tag{3.37}$$

A generator function can be defined also for the SO. It gives a simple way to determine the inverse SO (ISO) $S^{-1}(\cdot)$,

$$S^{-1}(v(t)) = w_I^T \underline{S}(v(t); a_I), \tag{3.38}$$

where w_I is the vector of weights and a_I the vector of thresholds of the ISO. The weights and the thresholds can be calculated from w and a by

$$a_{Ii} = \sum_{l=0}^{i} w_l (a_i - a_l), \ i = 0 \ldots L, \tag{3.39}$$

$$w_{I0} = \frac{1}{w_0}, \tag{3.40}$$

$$w_{Ii} - \frac{w_i}{\left(w_0 + \sum_{l=1}^{i} w_l\right)\left(w_0 + \sum_{l=1}^{i-1} w_l\right)}, \ i = 1 \ldots L, \tag{3.41}$$

$$a_{Ii} = \sum_{l=i}^{0} w_l (a_i - a_l), \ i = -L \ldots 0, \tag{3.42}$$

$$w_{li} = -\frac{w_i}{\left(w_0 + \sum_{l=i}^{-1} w_l\right)\left(w_0 + \sum_{l=i+1}^{-1} w_l\right)}, \quad i = -L ... -1, \tag{3.43}$$

where (3.39) and (3.41) are specific for the DZOs having a positive threshold, and (3.42)-(3.43) deal with the DZOs having a negative thresholds. Equation (3.40) defines the weight of the linear operator of the ISO, where obviously $a_{l0} = 0$.

3.3.6. The modified Prandtl-Ishlinskii model (MPIM)

The main limits of the PIM are the followings:

- *Odd symmetry* property: the hysteresis loops have an odd symmetry with respect to the center point of the loop (when the input oscillates between the same maximum and minimum values);
- *Convexity*: the strict monotonicity of the generator function given by (3.29) forces the generator function itself to be convex. As consequence, the PIM can describe hysteresis loops with increasing branches which are counter-clockwise oriented.

Two interesting approaches to overcome these limitations are suggested in [33] and [42]. The former, called the generalized PIM (GPIM), increses the flexibility of the ordinary PIM by considering *generalized hysterons* characterized by nonlinear functions. The latter consists in augmenting the overall hysteresis model with a SO which acts on the output $v(t)$ of the PIM $H[u; z_0]$. This augmented model is called the modified Prandtl-Ishlinskii model (MPIM). The MPIM will be denoted with $\Gamma[\cdot]$ and is schematized in Figure 3.9. As can be seen from Figure 3.9, the MPIM is essentially the series of the PIM, denoted by H, and the SO denoted by S.

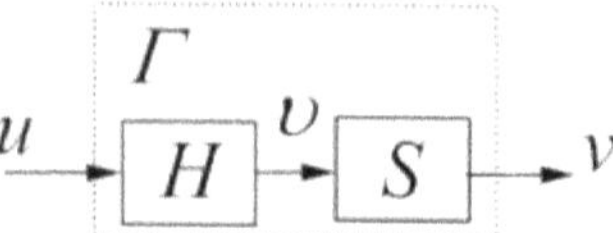

Figure 3.9 – Schema of the Modified Prandtl-Ishlinskii model (MPIM)

The MPIM is thus expressed by

$$v(t) = \Gamma\left[u;w,p,r,a\right](t) = S\left(H\left[u;r\right](t);a\right) = w^T \underline{S}\left(p^T \underline{H}\left[u;r\right](t);a\right). \tag{3.44}$$

The inverse MPIM (IMPIM) $\Gamma^{-1}[\cdot]$, is the series between the ISO and the IPIM and is expressed by

$$\begin{aligned} \Gamma^{-1}\left[v;w_I,p_I,r_I,a_I\right](t) &= H^{-1}\left[S^{-1}\left(v;a_I\right);r_I\right](t) = \\ &= p_I^T \underline{H}\left[w_I^T \underline{S}\left(v;a_I\right);r_I\right](t) \end{aligned}. \tag{3.45}$$

It can be noted that the MPIM is not LIP with respect to w and p. The identification requires the use of a particular trick discussed in section 3.3.7. Note that in (3.44) and (3.45), as well as in the remaining part of this thesis, the initial state vector z_0 is omitted since it is assumed $z_0 = 0$.

3.3.7. Identification of a MPIM

The identification has the goal of determining the parameters w, p, a, r of the MPIM (3.45), starting from the input signal $u(t)$ and output signal $v(t)$, for $t \in [t_0, t_E]$. We recall that the assumption of zero initial states is made in order to simplify the notation and the discussion, so $z_0 = 0$ and $z_{I0} = 0$.

The nonlinear dependence of the MPIM on its parameters can be avoided if an intelligent choice of the model error is done. This error e_1 is defined as (see Figure 3.10.)

$$e_1(w_I, p) = p^T \underline{H}[u;r] - w_I^T \underline{S}(v;a_I), \tag{3.46}$$

where the PIM and the ISO have been involved. This error is LIP in w_I and p. Consequently a quadratic optimization problem can be set up if a_I and r are fixed prior to the identification of w_I and p. Since $u(t)$ and output signal $v(t)$ are available, the threshold r can be calculated from (3.26), while the threshold can be calculated in a similar way by a uniform distribution on the output range, i.e.,

$$a_{Il} = \frac{-l}{L+1} \min_{t_0 \le t \le t_E} \{v(t)\}, \; l = -L..-1, \tag{3.47}$$

$$a_{Il} = \frac{l}{L+1} \max_{t_0 \le t \le t_E} \{v(t)\}, \; l = 0..L. \tag{3.48}$$

Note that, based on (3.26) and (3.48), the operators with $r_0 = 0$ and $a_0 = 0$ are linear. The functional to be minimized is defined to be [42]

$$F = \min\left\{ \begin{pmatrix} p^T & w_I^T \end{pmatrix} \cdot \int_0^{t_E} \begin{pmatrix} \underline{H}[u;r](t) \\ -\underline{S}(v(t);a_I) \end{pmatrix} \begin{pmatrix} \underline{H}[u;r](t)^T & -\underline{S}(v(t);a_I)^T \end{pmatrix} \mathrm{d}t \cdot \begin{pmatrix} p \\ w_I \end{pmatrix} \right\}, \tag{3.49}$$

subjected to the monotonicity constraints

$$\begin{cases} Pp \ge \delta_P \\ Ww_I \ge \delta_W \end{cases}, \tag{3.50}$$

where the matrix P expresses in a matrix notation (3.29), the matrix W expresses (3.37), $\delta_W = [\delta,..,\delta]^T \in \Re^{2L+1}$, $\delta_P = [\delta,..,\delta]^T \in \Re^{N+1}$, and $\delta > 0$ is a small positive constant which can be chosen by design. The mentioned matrices are defined by

$$P = \underline{T}_{N+1} \in \Re^{(N+1)\times(N+1)}, \tag{3.51}$$

where $\underline{T}_{N+1}$ is a low-triangular matrix with non-zero elements equal to 1, and

$$W = \begin{pmatrix} \underline{T}^L & \underline{1}^{L\times 1} & \underline{0}^{L\times L} \\ \underline{0}^{1\times L} & 1 & \underline{0}^{1\times L} \\ \underline{0}^{L\times L} & \underline{1}^{L\times 1} & \underline{T}_L \end{pmatrix} \in \Re^{(2L+1)\times(2L+1)}, \tag{3.52}$$

where $\underline{T}_n \in \Re^{n\times n}$ and $\underline{T}^n \in \Re^{n\times n}$ indicate, respectively, a low-triangular matrix with non-zero elements equal to 1 and a high-triangular matrix with the nonzero elements equal to 1; $\underline{1}^{n\times m} \in \Re^{n\times m}$ indicates a matrix with all elements equal to 1; $\underline{0}^{n\times m} \in \Re^{n\times m}$ is the zero matrix. It can be noted that also (3.49)-(3.50) is actually a constrained L_2^2-norm minimization problem.

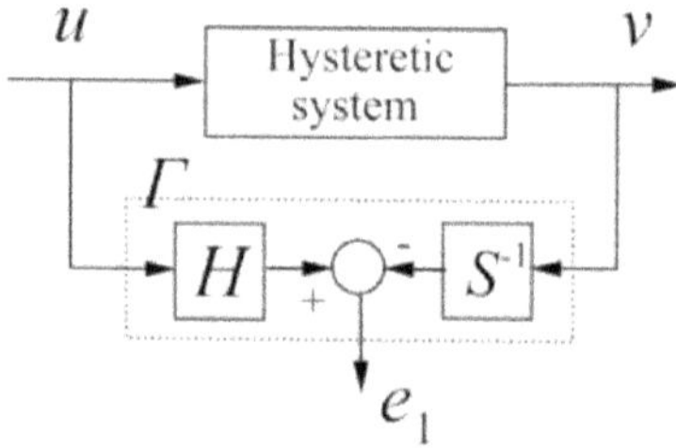

Figure 3.10 – Error model for the identification of the MPIM parameters

To summarize, the identification of a MPIM is done in the following steps:

I. Choice of the model orders L and N which define the number of elementary operators, respectively, of the SO ($2L+1$) and the PIM (N).

II. Distribution of the thresholds r of the PIM as in (3.26) on the input signal range.

III. Distribution of the negative thresholds a_l, with (3.47); distribution of the non-negative ones as in (3.48). The thresholds of the ISO are distributed on the output signal range.

IV. The initial states of the PIM z_0 and the ones of the IPIM z_{I0} are set to zero if the initialization procedure is assumed.

V. The quadratic optimization problem (3.49)-(3.50) is solved and the optimal values of w_I and p are found.

VI. Using (3.39)-(3.43) the parameters of the SO (w, a) are calculated.

3.4. Properties of hysteresis models

In sections 3.2 and 3.3 two models have been introduced, namely the KPM and the MPIM. Their inverses have been discussed. In this section, some properties of these two models are introduced. They are typically invoked in that part of the literature which deals with control of hysteretic systems by controllers that do not entail any compensation. For simplicity of notation, in the following the notation of a hysteresis operator with input u is simplified to $\Gamma[u]$ or just Γ, omitting the initial states and the other parameters as thresholds and weights. Obviously, once a particular model is chosen for Γ, the reader can recall the complete parameterization from sections 3.2 and 3.3.

Definition 3.6. *The numerical value set of a hysteresis operator Γ is the set $NVS(\Gamma)$ such that*

$$NVS(\Gamma)=\{\Gamma[u](t):u(t)\in C(\Re_+),\ \ t\in\Re_+\}. \tag{3.53}$$

The numerical value set does not have to be bounded, and this depends on the particular operator Γ.

Definition 3.7. *The operator Γ is said to be monotone if, for $u\in C(\Re_+)$, then*

$$\frac{\mathrm{d}\Gamma[u](t)}{\mathrm{d}u(t)}\frac{\mathrm{d}u(t)}{\mathrm{d}t}\geq 0\text{, a.e. for } t\in\Re_+ . \tag{3.54}$$

Definition 3.8. *The operator Γ is said to be globally Lipschitz continuous with constant L_C if for each $u_1,u_2\in C(\Re_+)$*

$$\sup_{\tau\in\Re_+}\left|\Gamma[u_1](\tau)-\Gamma[u_2](\tau)\right|\leq L_C\sup_{\tau\in\Re_+}\left|u_1(\tau)-u_2(\tau)\right|. \tag{3.55}$$

It can be noted that (3.55) is equivalent to (3.19) when $\tau \to +\infty$. Let v be the output of the hysteresis operator. If the hysteresis operator is Lipschitz continuous, it is a mapping between spaces of continuous functions, i.e., $\Gamma : u \in C(\Re_+) \to v \in C(\Re_+)$.

The Lipschitz constant defines how the variation on the output v is influenced by a certain variation of the input u. It is of interest the calculation of such constant.

3.4.1. Lipschitz constant of the KPM

Let $u_1, u_2 \in C(\Re_+)$ be the inputs to the KPM $\Gamma[\cdot]$; let $v_1, v_2 \in C(\Re_+)$ be the corresponding outputs $v_1 = \Gamma[u_1]$, $v_2 = \Gamma[u_2]$. The Lipschitz continuity of the KPM relies on the following property of the elementary KP operator,

$$\sup_{\tau \in \Re_+} \left| k_i[u_1](\tau) - k_i[u_2](\tau) \right| \leq \frac{2}{a} \sup_{\tau \in \Re_+} \left| u_1(\tau) - u_2(\tau) \right|, \tag{3.56}$$

which is then Lipschitz continuous with constant $2/a$.

***Proposition 3.1**. The KPM is globally Lipschitz continuous with constant*

$$L_C = \frac{2}{a} \sum_{i=1}^{K} |w_i|. \tag{3.57}$$

Proof. Since

$$\sup_{\tau \in \Re_+} \left| \Gamma[u_1](\tau) - \Gamma[u_2](\tau) \right| \leq \sum_{i=1}^{K} |w_i| \sup_{\tau \in \Re_+} \left| k_i[u_1](\tau) - k_i[u_2](\tau) \right|, \tag{3.58}$$

one obtains from (3.56) that

$$\sup_{\tau\in\Re_+}\left|\Gamma[u_1](\tau)-\Gamma[u_2](\tau)\right|\le\frac{2}{a}\sum_{i=1}^{K}\left|w_i\right|\sup_{\tau\in\Re_+}\left|u_1(\tau)-u_2(\tau)\right|, \quad (3.59)$$

which completes the proof.

3.4.2. Lipschitz constant of the MPIM

The calculation of the constant of the MPIM relies on two fundamental properties. Let $u_1,u_2\in C(\Re_+)$ be the inputs to the MPIM; let $v_1,v_2\in C(\Re_+)$ be the corresponding outputs $v_1=\Gamma[u_1]$, $v_2=\Gamma[u_2]$; let $\upsilon_1,\upsilon_2\in C(\Re_+)$ be the outputs of the PIM, respectively $\upsilon_1=H[u_1]$ and $\upsilon_2=H[u_2]$. For the elementary *j*-th play operator the following holds:

$$\sup_{\tau\in\Re_+}\left|H_j[u_1](\tau)-H_j[u_2](\tau)\right|\le\sup_{\tau\in\Re_+}\left|u_1(\tau)-u_2(\tau)\right|, \quad (3.60)$$

and for the elementary *l*-th DZO the following is verified

$$\sup_{\tau\in\Re_+}\left|S_l(\upsilon_1(\tau))-S_l(\upsilon_2(\tau))\right|\le\sup_{\tau\in\Re_+}\left|\upsilon_1(\tau)-\upsilon_2(\tau)\right|. \quad (3.61)$$

Proposition 3.2. *The MPIM Γ is Lipschitz continuous with constant*

$$L_C=\sum_{l=-L}^{L}\left|w_l\right|\sum_{j=0}^{N}\left|p_j\right|. \quad (3.62)$$

Proof. From (3.60) it comes that for the overall PIM it holds

$$\sup_{\tau\in\Re_+}\left|H[u_1](\tau)-H[u_2](\tau)\right|\le\sum_{j=0}^{N}\left|p_j\right|\sup_{\tau\in\Re_+}\left|u_1(\tau)-u_2(\tau)\right|, \quad (3.63)$$

while from (3.61) it is obtained that

$$\sup_{\tau\in\Re_+}\left|S(\upsilon_1(\tau)) - S(\upsilon_2(\tau))\right| \le \sum_{l=-L}^{L}|w_l|\sup_{\tau\in\Re_+}|\upsilon_1(\tau) - \upsilon_2(\tau)| \ . \tag{3.64}$$

From (3.63) and (3.64) one obtains

$$\sup_{\tau\in\Re_+}\left|\Gamma[u_1](\tau) - \Gamma[u_2](\tau)\right| \le \sum_{l=-L}^{L}|w_l|\sum_{j=0}^{N}|p_j|\sup_{\tau\in\Re_+}|u_1(\tau) - u_2(\tau)|, \tag{3.65}$$

and thus the proof is completed.

4. PID control

Several applications require the precise positioning of a load by means of the movement of an end-effector. There are basically two types of positioning systems. Some of them require that the end-effector moves by steps to reach the desired position. Other systems require instead that the end-effector moves following sinusoidals or triangular waves. In the first case the objective of the controller is the tracking of constant references; in second case the objective is the tracking of time-varying references.

Standard controllers (PIs and PIDs) have shown to be very effective for the tracking of constant references in a large variety of situations and applications. The focus of this chapter is the investigation of PID control in the case of hysteretic systems, with or without a linear dynamic part.

In particular, a new approach to assess the asymptotic stability of the resulting closed-loops is proposed. Section 4.1 presents the control problem and the kind of closed-loop that is considered in this chapter.

Section 4.2 presents the new control approach. First, a way to express the derivative of a hysteresis output is proposed. Then, the expression is used to introduce the approach for high order systems and low order systems, where the order refers to the dynamic part. Section 4.2.7 discusses a detailed comparison between the proposed control approach and existing results.

Section 4.3 presents some experimental results obtained by applying the proposed method. The results have been acquired on two different MSM spring actuators, enforcing the validity of the approach.

An important assumption that is done in this chapter is that the hysteresis is time-invariant. For MSM actuators, this holds true in the case that (1) the actuator design is particularly robust with respect to temperature variations, or (2) those variations are neglectable, or (3) the hysteresis is not so sensitive to temperature in some temperature ranges (which is the case for some MSM materials). Moreover, the chapter focuses on the tracking of constant references. There are no results in literature for the tracking of other references by means of PID controllers. A way to overcome the two mentioned limitations is proposed in Chapter 5.

4.1. Statement of the control problem

4.1.1. The control-loop

The control-loop that is analyzed in this chapter is sketched in Figure 4.1.

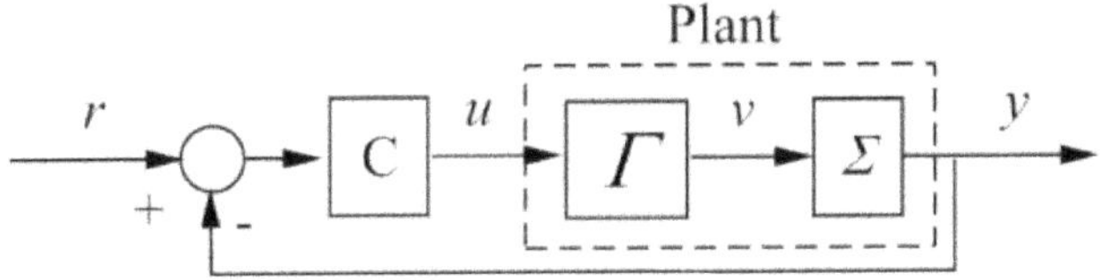

Figure 4.1 – Control-loop for dynamic systems Σ with an input hysteresis Γ

The plant indicates the presence of a smart-material based actuator or, in general, of a hysteretic dynamic plant that can be described as in Figure 2.9. The hysteresis model is denoted by Γ, and can be a MPIM or a KPM. The linear dynamic part, denoted by Σ, is assumed to be described by

$$\Sigma : \begin{cases} \dot{x} = Fx + Gv \\ y = x_1 \\ x(t_0) = x(0) \end{cases}, \tag{4.1}$$

where $x = [x_1, .., x_n]^T \in \Re^n$,

$$F = \begin{bmatrix} 0 & 1 & 0 & \dots & 0 \\ 0 & 0 & 1 & \dots & 0 \\ \vdots & \vdots & \vdots & \ddots & \vdots \\ 0 & 0 & 0 & \dots & 1 \\ a_{n1} & a_{n2} & a_{n3} & \dots & a_{nn} \end{bmatrix} \in \Re^{n \times n}, \tag{4.2}$$

$$G = \begin{bmatrix} 0 \\ \vdots \\ 0 \\ b \end{bmatrix} \in \Re^n , \tag{4.3}$$

$v \in \Re$ and $u \in \Re$. The output of the linear dynamics (and the overall plant) y is the measured position to be used in the feedback. The reference r is constant, such that the derivatives $r^{(i)} = 0$ for $i \geq 1$. This is a reasonable assumption for a certain class of positioning systems.

The controller is denoted by C and is a standard controller with a PID structure, i.e., the control output u can be expressed by

$$u = -K_P(x_1 - r) - K_I \int_0^t (x_1 - r)\mathrm{d}\tau - K_D \frac{\mathrm{d}}{\mathrm{d}t}(x_1 - r), \tag{4.4}$$

where the parameters K_P, K_I and K_D are the gains of the controller. The error $x_1 - r$ is called the *tracking error*. Note that the output of the hysteresis is denoted by v: the precise expression depends on the hysteresis model. In the case of a MPIM, v is expressed by (3.44); in the case of KPM, v is in (3.11)-(3.12).

A first question concerns the existence and uniqueness of the solution of (4.1). It is possible to state the following.

Proposition 4.1. *The initial value problem (4.1), where* $v = \Gamma[u]$ *and* Γ *is a KPM (3.11)-(3.12) or a MPIM (3.44), has a unique solution* $x(t) \in C^1[0, +\infty)$.

Proof. It is based on standard arguments [43] which make use of the global Lipschitz continuity of the KPM and the MPIM (see section 3.4).

Although the control-loop in Figure 4.1 is the focus of our discussion, literature often considers the alternative description depicted in Figure 4.2, where the hysteresis is said to appear in the "damping path". This representation is used to model nonlinear spring-like effects or friction phenomena [44].

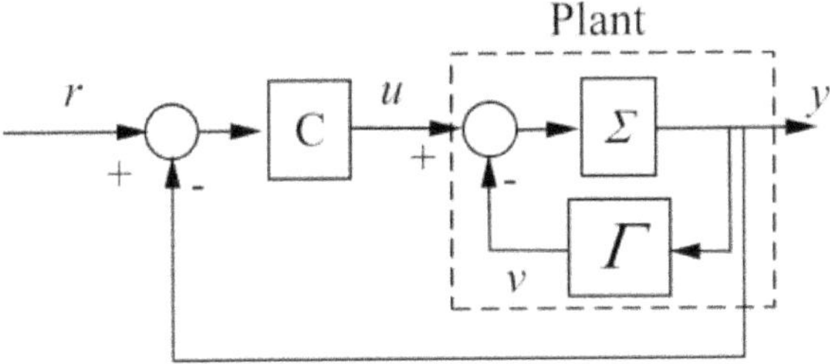

Figure 4.2 – Control-loop for dynamic systems with a hysteresis in the damping path

The goal of the standard controller C is to make the measured position follow the reference constant trajectory r, such that $x_1 \to r$ as $t \to +\infty$. There are several issues which are of interest. Is it possible to ensure the asymptotic convergence of the tracking error with a standard PID? If so, how fast is the convergence? Can the convergence behavior be shaped by the designer? The method proposed in the next pages tries to answer to those questions.

4.1.2. Literature review

Due to the simplicity of standard controllers, recently the topic of PID control of systems with hysteresis has attracted much research. This section aims at giving a brief panoramic view about the existing analytical results. Only the main aspects are emphasized while the technical details are neglected. A critical comparison between the approach proposed in this thesis and those ones will take place later.

Concerning smart-material-based systems, reference [45] investigates the stability property of the loop in Figure 4.2, for a particular case where the hysteresis model is a Bouc-Wen model. This model is quite flexible, depends on just five parameters, and has been widely used in the scientific community, especially to model the nonlinear spring-like effects of mechanical structures. The analysis of [45] carries out some bounds on the PID controller parameters, which ensure the stability of the control-loop. Reference inputs that are constant and nonzero are not handled.

Reference [46] deals with pure integral control of hysteretic dynamic systems with input and output nonlinearities. The integral gain is time-varying. The work is able to address most of the hysteresis models (for instance, the MPIM and the KPM), since it only needs to characterize the hysteresis in terms of its Lipschitz-constant. A set of assumptions on the hysteresis, involving the monotonicity and the local/global Lipchitz continuity, must be satisfied.

The paper from Jayawardhana et al. [41] specializes the theoretical framework of [46] for the case of PID control of hysteretic systems with a second-order dynamics. The interesting feature of the paper is that it considers both topologies of Figure 4.1 and Figure 4.2. We summarize here the result that concerns the first topology. Let Σ be characterized in terms of the damping coefficient c, the mass m and the stiffness k,

$$\Sigma : \begin{cases} \dot{x}_1 = x_2 \\ \dot{x}_2 = -\dfrac{k}{m}x_1 - \dfrac{c}{m}x_2 + \dfrac{1}{m}\Gamma[u] \\ y = x_1 \\ x_1(0);\, x_2(0) \end{cases}, \qquad (4.5)$$

and let the control action u be defined as in (4.4).

Theorem 4.1 [41]. *Let Γ be a hysteresis operator which is monotone, globally Lipschitz continuous with constant $L_C > 0$. If the gains of the PI controller are chosen satisfying*

$$0 < K_I < \frac{K_P k}{c}, \qquad (4.6)$$

$$\frac{K_I c}{k} < K_P < \frac{c^2}{L_C m}, \tag{4.7}$$

and, if the gains of the PID controller are chosen such that

$$0 < K_I < +\infty, \tag{4.8}$$

$$K_P > \frac{cK_I}{k}, \tag{4.9}$$

$$K_D > \frac{mK_P}{c}, \tag{4.10}$$

then there exist a unique solution $y(t) \in C^2[0,+\infty)$ *of the closed-loop and the output* $y(t)$ *converges asymptotically to the constant reference r.*

It can be noted that the controller gains are limited by the Lipschitz constant of the hysteresis model in the case of a PI controller. In the case of PID control, the gains are instead lower bounded by the choice of the integral gain. An application of theorem 4.1 to control a MSM push-push actuator can be found in [29].

The fact that the gains are limited by the Lipschitz constant returns in the paper [47], where the PI control of purely hysteretic systems is addressed. The considered loop is the one in Figure 4.1, which in this case is expressed by

$$\Sigma : \begin{cases} x_1 = \Gamma[u] \\ y = x_1 \end{cases}. \tag{4.11}$$

The result of the paper can be summarized as follows.

Theorem 4.2 [47]. *Let* Γ *be a hysteresis operator which is monotone and globally Lipschitz continuous with constant* $L_C > 0$. *If the gains of the PI controller satisfy*

$$K_I > 0 , \tag{4.12}$$

$$0 \le K_P < \frac{1}{L_C}, \tag{4.13}$$

then the closed-loop has a unique solution $y(t) \in C[0,+\infty)$ *and the tracking error* $y - r$ *decreases monotonically and asymptotically. Moreover, the BIBO stability is ensured also if r is not constant.*

The two results cannot be compared since they refer to different situations. It is possible to emphasize the following aspects.

a. Theorems 4.1 and 4.2 offer the bounds on the controller gains which ensure the asymptotic tracking of a constant reference.
b. Theorems 4.1 and 4.2 do not determine the type of convergence of the tracking error. Theorem 4.2 only specifies that it decreases monotonically, thus overshoots are not possible.
c. Theorem 4.1 does not handle references which are not constant, whereas theorem 4.2 proves the BIBO stability of the loop in the case of arbitrary references.
d. Theorems 4.1 and 4.2 do not define a way to find the best controller gains. In other words, there are no performance indexes that can be related to the specific controller.
e. The results of theorems 4.1 and 4.2 cannot be extended to higher order system.
f. Both theorems handle easily the presence of model uncertainties.
g. Theorems 4.1 and 4.2 give conditions for the stability and the tracking that are only sufficient.

In section 4.2 an alternative method to determine the asymptotic tracking of control-loops with hysteresis is proposed. A detailed comparative discussion with theorems 4.1 and 4.2 follows in section 4.2.7.

4.2. A new approach for PID control

The main feature of the new approach detailed in this section is that it transforms the tracking problem of the nonlinear system (4.1) with hysteresis in the stability problem of a linear perturbed system [48] [49], allowing the designer to take advantage of widely-used design methods based on the Linear Matrix Inequality framework [49]. Three cases are here treated separately. They differ in the calculations but not in the general framework. The first concerns the control of systems whose order is $n \geq 2$. Here the main concepts are introduced. The remaining cases $n=0$ and $n=1$ are addressed successively. After, the case $n=2$ is discussed in more details because of its importance in the context of smart-material-based actuators. However, it is necessary to first focus on a particular issue that is relevant for the forthcoming developments, and this is done in section 4.2.1.

4.2.1. Derivative of the hysteresis output

The goal of this section is to find a convenient expression of the time derivative of the output of a MPIM or a KPM (refer to section 3.2.2 and 3.3.6 for more details about the two models). Let us first focus on the MPIM, and let u be the input to the hysteresis and v be the output.

Recall the parameterization (3.44), from which it is obtained

$$\frac{\mathrm{d}}{\mathrm{d}t}v = \frac{\mathrm{d}}{\mathrm{d}t}\Gamma[u] = \frac{\mathrm{d}}{\mathrm{d}t}\left(\sum_{l=-L}^{L} w_l S_l\left(\sum_{j=0}^{N} p_j H_j[u]\right)\right). \tag{4.14}$$

The reader should now recall that every elementary operator, DZO or play, gives a constant output for some input ranges and internal state conditions. This can be visually appreciated in Figure 3.7 and Figure 3.8. These input ranges are determined by the threshold of the elementary operator. However, to compute the derivative in (4.14), only those DZOs and those play operators which have a non-constant output are relevant. Let us call those operators as *active*.

To consider those operators in a formal way, let us define the *active sets*

$$\Omega_j^A(t) = \left\{ j \in [0, N] \mid \frac{\mathrm{d}}{\mathrm{d}t} H_j[u] \neq 0 \right\}, \tag{4.15}$$

$$\Omega_l^A(t) = \left\{ l \in [-L, L] \mid \frac{\mathrm{d}}{\mathrm{d}t} S_l(v) \neq 0 \right\}, \tag{4.16}$$

which trace at each moment t the DZOs and play operators that are active. With (4.15) and (4.16), the derivative in (4.14) can be written as

$$\frac{\mathrm{d}}{\mathrm{d}t} \Gamma[u] = \sigma(t) \frac{\mathrm{d}}{\mathrm{d}t} u \,, \tag{4.17}$$

where

$$\sigma(t) = \sum_{l \in \Omega_l^A(t)} w_l \sum_{j \in \Omega_j^A(t)} p_j \,. \tag{4.18}$$

The variable $\sigma(t)$ changes each time that a new operator, DZO or play, becomes active or is deactivated. Since the number of operators is finite, and since the characteristics of the DZO and the play are differentiable almost everywhere, the values of $\sigma(t)$ belong to a finite set of values and jump between each value of this set depending on the input u and its history. A maximum and minimum value of $\sigma(t)$ can be defined,

$$\sigma_- = \min_{\substack{l \in -L..L \\ j \in 0..N}} \{\sigma(t)\} = \min \left\{ \min_{\substack{l \in 0..L \\ j \in 0..N}} \{\sigma(t)\}, \min_{\substack{l \in -L..0 \\ j \in 0..N}} \{\sigma(t)\} \right\}, \tag{4.19}$$

$$\sigma^{+}=\max_{\substack{l\in -L..L\\ j\in 0..N}}\{\sigma(t)\}=\max\left\{\max_{\substack{l\in 0..L\\ j\in 0..N}}\{\sigma(t)\},\max_{\substack{l\in -L..0\\ j\in 0..N}}\{\sigma(t)\}\right\}, \tag{4.20}$$

such that $\sigma(t)\in[\sigma_{-},\sigma^{+}]$ for all $t\in\Re^{+}$.

It is worth noting that, if the MPIM is strictly monotone, then $\sigma_{-}>0$ and consequently $\sigma^{+}>0$. The strict monotonicity will be shown to be a requirement of the control approach that is proposed.

Let us now focus on the KPM. The discussion made for the MPIM is very similar. Thanks to the fact that only the active KP operators give a contribution to the derivative, it is obtained the active set

$$\Omega_i^A(t)=\left\{i\in[1,K]\,|\,\frac{\mathrm{d}}{\mathrm{d}t}k_i[u]\neq 0\right\}, \tag{4.21}$$

and the output derivative can be expressed again as in (4.17), with the new definition

$$\sigma(t)=\frac{2}{a}\sum_{i\in\Omega_i^A(t)}w_i\,. \tag{4.22}$$

The variable $\sigma(t)$ jumps between a maximum and a minimum value, respectively

$$\sigma_{-}=\min_{i\in 1..K}\{\sigma(t)\}, \tag{4.23}$$

$$\sigma^{+}=\min_{i\in 1..K}\{\sigma(t)\}. \tag{4.24}$$

The reason why the same symbol $\sigma(t)$ has been used for both KMP and MPIM is that a unified discussion will be done for both models.

But...what is the meaning of $\sigma(t)$? It sums the weights of each elementary operator, for a KPM or a MPIM. If the MPIM is thermodynamically consistent [42], then each weight is bigger than zero and $\sigma_- > 0$ (consequently $\sigma^+ > 0$). The KPM is generally identified by forcing the weights to be non-negative due to invertibility and continuity issues. In this case, $\sigma_- > 0$ and thus $\sigma^+ > 0$. The meaning of σ^+ is more evident if one relates it to the Lipschitz constant of the hysteresis model. In the case of a KPM with non-negative weights, σ^+ is *equivalent* to the Lipschitz constant in (3.57). In the case of a MPIM which is thermodynamically consistent, σ^+ is *equivalent* to the Lipschitz constant in (3.62). If the MPIM is only monotone and not thermodynamically consistent, then $0 < \sigma^+ \leq L_C$ since some weights can be negative. The relationship between σ^+ and the Lipschitz constant of the models facilitates the comparisons between the control approach proposed here and existing results which are available in literature.

4.2.2. Control of high order systems (*n=2 or n>2*)

Let us define the vector of the desired trajectories

$$x_D = \left[\int_0^t r \mathrm{d}\tau, r, \dot{r}, .., r^{(n-1)} \right]^T \in \Re^{n+1}, \tag{4.25}$$

which also entails the integral of the desired reference *r*. Let us define the augmented state vector

$$x_A = [x_0, x]^T \in \Re^{n+1}, \tag{4.26}$$

where $x_0 = \int_0^t x_1 \mathrm{d}\tau$. Let us introduce the vector of the tracking error and its derivatives (and integral)

$$e = x_A - x_D = [e_0, e_1, .., e_n]^T \in \Re^{n+1}, \tag{4.27}$$

where $e_0 = \int_0^t e_1 \mathrm{d}\tau$, $e_1 = x_1 - r$, $e_i = x_i - r^{(i-1)}$ for $i = 2..n$. The dynamics of the tracking error vector are described by

$$\begin{cases} \dot{e} = \dot{x}_A - \dot{x}_D = \bar{F}x_A + \bar{G}v - \dot{x}_D \\ e(0) = x_A(0) - x_D(0) \end{cases}, \tag{4.28}$$

where the matrices $\bar{F}$ and $\bar{G}$ are defined as

$$\bar{F} = \begin{bmatrix} 0 & 1 & 0 & \cdots & 0 \\ 0 & & & & \\ \vdots & & F & & \\ 0 & & & & \end{bmatrix} \in \Re^{(n+1)\times(n+1)}, \tag{4.29}$$

$$\bar{G} = \begin{bmatrix} 0 \\ G \end{bmatrix} \in \Re^{(n+1)\times 1}. \tag{4.30}$$

It is convenient considering a new state vector

$$\xi = \dot{e} = [\xi_0, \xi_1, ..., \xi_n]^T \in \Re^{n+1}, \tag{4.31}$$

where it is pointed out that $\xi_0 = e_1$,i.e., the tracking error that the controller has to handle. The dynamics of ξ are determined by calculating the time derivative of (4.28) and substituting the definition of e in (4.27),

$$\begin{cases} \dot{\xi} = \bar{F}\xi + \bar{G}\left(\frac{\mathrm{d}}{\mathrm{d}t}v\right) + \bar{F}\dot{x}_D - \ddot{x}_D \\ \xi(0) = \dot{e}(0) \end{cases}. \tag{4.32}$$

When the reference is constant, as is the case treated here, then $\bar{F}\dot{x}_D - \ddot{x}_D = 0$, and the dynamics of ξ reduce to

$$\dot{\xi} = \bar{F}\xi + \bar{G}\dot{v} . \tag{4.33}$$

It can be noted that in (4.33) the time derivative of the hysteresis output, $\dot{v}$, appears. The reader can refer to (4.17)-(4.18) if the hysteresis model is a MPIM, or to (4.17)-(4.22) if it is a KPM. Due to the definition of u in (4.4) and ξ in (4.31), the derivative (4.17) becomes

$$\frac{\mathrm{d}}{\mathrm{d}t}\Gamma[u] = \sigma(t)\left(-K_P\xi_1 - K_I\xi_0 - K_D\xi_2\right) . \tag{4.34}$$

Thanks to (4.34), the closed-loop (4.32) becomes

$$\dot{\xi} = A\left(\sigma(t)\right)\xi , \tag{4.35}$$

where $A(\sigma(t))$ is the same as $\bar{F}$ except for the first three elements of the last row,

$$A\left(\sigma(t)\right) = \begin{bmatrix} 0 & 1 & 0 & \dots & 0 \\ 0 & 0 & 1 & \dots & 0 \\ \vdots & \vdots & \vdots & \ddots & \vdots \\ 0 & 0 & 0 & \dots & 1 \\ \bar{a}_{n0} & \bar{a}_{n1} & \bar{a}_{n2} & \dots & \bar{a}_{nn} \end{bmatrix} \in \Re^{(n+1)\times(n+1)} , \tag{4.36}$$

and $\bar{a}_{n0} = -bK_I\sigma(t)$, $\bar{a}_{n1} = \left(a_{n1} - bK_p\sigma(t)\right)$, $\bar{a}_{n2} = \left(a_{n2} - bK_D\sigma(t)\right)$, $\bar{a}_{ni} = a_{ni}$ for $i = 3..n$. Let us define the matrices

$$\begin{aligned} A_- &= A\left(\sigma(t) = \sigma_-\right) \\ A_+ &= A\left(\sigma(t) = \sigma^+\right) \end{aligned} . \tag{4.37}$$

It is evident that the behavior of the tracking error ξ_0 depends on the stability of the system (4.35) with the state matrix (4.36). The following two propositions are verified.

Proposition 4.2. *The closed-loop (4.35) has a unique solution* $\xi(t) \in C[0,+\infty)$ *independently of the controller gains.*

Proof. Refer to the matrices (4.37). When $\sigma(t)$ varies, the matrix $A(\sigma(t))$ changes monotonically between A_- and A^+, both having a bounded norm. Consequently, any norm of $A(\sigma(t))$ is bounded, i.e., $\|A(\sigma(t))\| \leq \|A_+\|$. This means that $A(\sigma(t))\xi$ is Lipschitz continuous in ξ and piecewise continuous in *t*. Note that the Lipschitz condition holds globally for each $t \in \Re^+$. By standard arguments, the solution $\xi(t)$ must be unique, $\xi(t) \in C[0,+\infty)$, and piecewise differentiable. This conclusion is independent of the controller gains.

Proposition 4.3. *The matrix* $A(\sigma(t))$ *is the convex combination of the matrices* A_- *and* A_+ *in (4.37), i.e., for all* $t \in \Re^+$ *there exists a positive coefficient* $\lambda(t)$, $0 \leq \lambda(t) \leq 1$, *such that*

$$A(\sigma(t)) = \lambda(t) A_+ + (1 - \lambda(t)) A_- . \qquad (4.38)$$

Proof. The relationship (4.38) must hold for each element of the matrix $A(\sigma(t))$ in order for (4.38) to be true. It is trivially satisfied for all elements, except for the first three elements of the last row, $\overline{a}_{n0}$, $\overline{a}_{n1}$ and $\overline{a}_{n2}$. Note that, since $\sigma(t) \in [\sigma_-, \sigma^+]$, it is always possible to find a $\lambda(t)$, $0 \leq \lambda(t) \leq 1$, which verifies

$$\sigma(t) = \lambda(t)\sigma^+ + (1 - \lambda(t))\sigma_- . \qquad (4.39)$$

Let us focus on $\overline{a}_{n0}$. Substituting (4.39) into the expression of $\overline{a}_{n0}$, one obtains

$$-bK_I\sigma(t) = \lambda(t)\left(-bK_I\sigma^+\right) + \left(1-\lambda(t)\right)\left(-bK_I\sigma_-\right), \tag{4.40}$$

which actually shows that the term $\overline{a}_{n0}$ is the convex combination of the corresponding terms in A_- and A_+. A similar discussion is valid for $\overline{a}_{n1}$ and $\overline{a}_{n2}$. Hence, the proof is completed.

The convexity property of $A(\sigma(t))$ is extremely useful to assess the asymptotic stability of (4.35). In fact, thanks to (4.38), the system (4.35) can be thought as a *polytopic linear differential inclusion* (PLDI) [50], [51]. A very interesting result concerning the stability of such systems is reported next, specialized for the case under discussion.

Theorem 4.2 [51]. *The system (4.35) is exponentially stable if the matrices* A_- *and* A_+ *share a common quadratic Lyapunov function, i.e., if a positive definite matrix P exists such that*

$$A_-^T P + PA_- < 0 \text{ and } A_+^T P + PA_+ < 0. \tag{4.41}$$

The existence of *P* is equivalent to the existence of a pair (P, α) ,with $\alpha > 0$, that verifies [51]

$$\xi^T(t)P\xi(t) \le \xi^T(0)P\xi(0)e^{-\alpha t}, \tag{4.42}$$

and thus

$$\left\|\xi(t)\right\|_P \le \left\|\xi(0)\right\|_P e^{-\alpha t}, \tag{4.43}$$

where $\left\|\xi(t)\right\|_P = \sqrt{\xi^T(t)P\xi(t)}$. Equation (4.43) tells that the state vector ξ decreases exponentially fast, at a rate which is given by α. Note that both matrices A_- and

A_+ *must* admit an own quadratic Lyapunov function in order for such a P to exist, and thus they *must* be asymptotically stable on their own.

Theorem 4.2 offers a way to check whether some chosen gains K_P, K_I and K_D, which give contribution to the state matrix $A(\sigma(t))$, lead to an asymptotically stable closed-loop. Unfortunately, the theorem does not give a way to calculate the gains. It is however interesting to note that, since the gains lead to a certain P, they also lead to the corresponding α, which can represent a *performance index* of the closed-loop.

The *minimum decay rate* α is defined as the biggest positive α such that

$$\lim_{t\to+\infty} e^{\alpha t}\left\|\xi(t)\right\| = 0 \tag{4.44}$$

holds for every trajectory of (4.35). It can be determined by solving the optimization problem

$$\begin{aligned}&\text{maximize } \alpha\\ &\text{subject to } \begin{cases} A_-^T P + PA_- + 2\alpha P < 0 \\ A_+^T P + PA_+ + 2\alpha P < 0 \end{cases}.\end{aligned} \tag{4.45}$$

The minimum decay rate is a "safe" evaluation of the closed-loop: the controlled variable x_1 will reach the reference r in a maximum time related to α. It will be shown in section 4.3 that in our case the minimum decay rate provides an excessive underestimate of the real decay rate.

4.2.3. PI control of purely hysteretic systems (*n=0*)

Section 4.2.2 introduced the main idea behind the approach for PID control proposed in this thesis. It concerns the reformulation of the original nonlinear control problem (4.1) into the stability problem of a linear perturbed system (4.35). The same idea can be applied to purely hysteretic systems.

The dynamic part of the system now reads

$$\Sigma : \begin{cases} x_1 = \Gamma[u] \\ y = x_1 \end{cases}, \tag{4.46}$$

where $x_1, y \in \Re$, and $u \in \Re$ is defined as in (4.4). By applying the same mathematical steps illustrated before, it is obtained

$$\xi_1 = \sigma(t)\left(-K_P\xi_1 - K_I\xi_0\right), \tag{4.47}$$

which leads to

$$\dot{\xi}_0 = A\left(\sigma(t)\right)\xi_0, \tag{4.48}$$

where

$$A\left(\sigma(t)\right) = -\frac{K_I\sigma(t)}{1 + K_P\sigma(t)}. \tag{4.49}$$

Note that in this case $A\left(\sigma(t)\right)$ is a scalar quantity.

Proposition 4.4. *The system (4.48) admits a unique solution* $\xi_0(t) \in C[0, +\infty)$ *independently of the controller gains.*

Proof. It is based on the same arguments of proposition 4.2. In fact, $A\left(\sigma(t)\right)$ is strictly monotone and has a maximum value A_+ for $\sigma(t) = \sigma^+$ and a minimum value A_- for $\sigma(t) = \sigma_-$. Thus, any norm of $A\left(\sigma(t)\right)$ is bounded, and $A\left(\sigma(t)\right)\xi_0$ is Lipschitz continuous in ξ_0.

The convexity property of $A(\sigma(t))$ with respect to A_+ and A_- as in (4.37) also holds. Therefore, theorem 4.3 can be used to determine the asymptotic stability of (4.48). It can be noted that in this case the hypothesis of theorem 4.1 are *always* satisfied, for each $P>0$, $P\in\Re$, if

$$K_I > 0, \tag{4.50}$$

$$K_P > -\frac{1}{\sigma^+}, \tag{4.51}$$

In other words, the system (4.48) is asymptotically stable for any positive value of the controller gains. The minimum decay rate can be estimated by using (4.45). Recall the relationship between σ^+ and the Lipschitz constant of the hysteresis models. The limitation of the proportional gain is related to the inverse of this constant.

4.2.4. PI control of first-order systems (*n=1*)

The dynamic part of the system now reads

$$\Sigma:\begin{cases}\dot{x}_1 = -\chi x_1 + \beta\Gamma[u] \\ y = x_1\end{cases}, \tag{4.52}$$

where $\beta,\chi\in\Re$. Note that if $\chi>0$ then the linear system (4.52) is asymptotically stable. β is the input gain, and in the following it is assumed that $\beta>0$. With similar manipulations as before, it is obtained a dynamic system of the error vector $\xi=[\xi_0,\xi_1]^T$ of the form (4.35) with

$$A(\sigma(t)) = \begin{bmatrix} 0 & 1 \\ -\beta K_I\sigma(t) & -(\chi+\beta K_P\sigma(t)) \end{bmatrix}. \tag{4.53}$$

The existence and uniqueness of the solution $\xi(t)\in C[0,+\infty)$, and the convexity of $A(\sigma(t))$ with $A_- = A(\sigma=\sigma_-)$ and $A_+ = A(\sigma=\sigma^+)$, are verified. Thus, theorem 4.1 can be used to assess the asymptotic stability of (4.35)-(4.53). There are however further conditions that can be exploited to limit the range where the controller gains have to be searched. In fact, as mentioned in section 4.2.2, in order for the matrix P to exist each linear time-invariant system determined by A_- and A_+ must be asymptotically stable. In mathematical terms, the matrices

$$A_- = \begin{bmatrix} 0 & 1 \\ -\beta K_I \sigma_- & -(\chi+\beta K_P \sigma_-) \end{bmatrix}, \tag{4.54}$$

$$A_+ = \begin{bmatrix} 0 & 1 \\ -\beta K_I \sigma^+ & -(\chi+\beta K_P \sigma^+) \end{bmatrix}, \tag{4.55}$$

must have their eigenvalues in the left side of the complex plane. The Routh-Hurwitz criterion states that the matrices (4.54) and (4.55) are asymptotically stable if the following conditions hold:

$$\beta K_I \sigma_- > 0 \text{ and } \beta K_I \sigma^+ > 0, \tag{4.56}$$

$$(\chi+\beta K_P \sigma_-) > 0 \text{ and } (\chi+\beta K_P \sigma^+) > 0. \tag{4.57}$$

Constraints (4.56) are satisfied if $\sigma_-, \sigma^+ > 0$ and

$$K_I > 0, \tag{4.58}$$

the latter becoming a design requirement. Constraints (4.57) are satisfied if

$$K_P > -\frac{\chi}{\beta\sigma^+}, \tag{4.59}$$

which shows that the proportional gain is lower bounded by a (negative) quantity that is inversely proportional to σ^+ and directly proportional to the ratio χ / β. Note that, differently from the case of a purely hysteretic system, the bounds (4.58) -(4.59) are only *necessary* for the matrix P of theorem 4.3 to exist. Condition (4.41) must be checked to guarantee that the chosen gains ensure the asymptotic tracking.

4.2.5. PID control of second order systems (*n=2*)

This case is detailed due to its importance in the context of smart-material-based actuators [49]. They are often modeled as the series of a hysteresis and a second-order mechanical dynamics (see Figure 2.9). The latter is characterized by the damping coefficient *c*, the mass *m* and the stiffness *k*, in this way

$$\Sigma : \begin{cases} \dot{x}_1 = x_2 \\ \dot{x}_2 = -\dfrac{k}{m}x_1 - \dfrac{c}{m}x_2 + \dfrac{1}{m}\Gamma[u] \\ y = x_1 \\ x_1(0);\, x_2(0) \end{cases} . \tag{4.60}$$

Following the same mathematical steps as before, it is obtained the dynamic system (4.35) where $\xi = [\xi_0, \xi_1, \xi_2]^T$ and

$$A(\sigma(t)) = \begin{bmatrix} 0 & 1 & 0 \\ 0 & 0 & 1 \\ -\dfrac{K_I\sigma(t)}{m} & -\dfrac{(k + K_P\sigma(t))}{m} & -\dfrac{(c + K_D\sigma(t))}{m} \end{bmatrix} . \tag{4.61}$$

The matrix $A(\sigma(t))$ is again the convex combination of A_- and A_+ in (4.37). Theorem 4.1 gives the way to check the asymptotic stability of (4.35) with (4.61).

In order for a matrix P to exist, A_- and A_+ must be asymptotically stable. The application of the Routh-Hurwitz criterion [36] leads to the following constraints

$$c + K_D\sigma_- > 0 \text{ and } c + K_D\sigma_+ > 0, \tag{4.62}$$

$$K_I\sigma_- > 0 \text{ and } K_I\sigma^+ > 0, \tag{4.63}$$

$$\begin{cases} K_I < \dfrac{(c + K_D\sigma_-)(k + K_P\sigma_-)}{m\sigma_-} \\ K_I < \dfrac{(c + K_D\sigma^+)(k + K_P\sigma^+)}{m\sigma^+} \end{cases}, \tag{4.64}$$

which become as follows

$$K_D > -\frac{c}{\sigma_-}, \tag{4.65}$$

$$0 < K_I < \frac{(c + K_D\sigma_-)(k + K_P\sigma_-)}{m\sigma^+}. \tag{4.66}$$

The inequality (4.66) forces the following bound for the proportional gain

$$K_P > -\frac{k}{\sigma_-}. \tag{4.67}$$

It can be noted that the verification of (4.63) requires that the hysteresis model is strictly monotone in the case of a MPIM, or invertible in the case of a KPM, and that $K_I > 0$. These are thus requirements of the proposed method.

4.2.6. Discussion

In this section some interesting issues are analyzed. A first one concerns the behavior of the integral control action during the tracking. A second one concerns what happens to the closed-loop if the reference signal is not constant in some time intervals. The third issue is about the robustness of the control algorithm with respect to model uncertainties. Finally, a better performance index of the closed-loop is introduced.

Proposition 4.5. *Let the controller gains be chosen such that theorem 4.3 holds. Let r be a constant reference. The integral control action converges asymptotically to a finite value, i.e.,* $\lim_{t\to+\infty} K_I \int_0^t \xi_0(t)\mathrm{d}\tau < \infty$.

Proof. It has been shown that the trajectories of (4.35) converge asymptotically to zero at an exponential rate which is at least α from (4.45). This is expressed by (4.43), which can be rewritten as

$$\|\xi(t)\| \le e^{-\alpha t}\kappa(P)\|\xi(0)\|, \tag{4.68}$$

where $\kappa(P)$ is the condition number of P. Integrating both sides of (4.68) it is obtained

$$\int_0^{+\infty} \|\xi(t)\|\mathrm{d}t < +\infty, \tag{4.69}$$

which means that the function $\|\xi(t)\|$ belongs to the Banach space $L_1[0,+\infty)$. From the definition of ξ in (4.31) and (4.69) it is obtained that

$$\|e(t)\| \le \int_0^{+\infty} \|\xi(t)\|\mathrm{d}t \le +\infty, \tag{4.70}$$

and thus $e(t) \in L_\infty[0,+\infty)$. Note that $e(t)$ contains the integral term e_0, and the proposition is proved.

Theorem 4.4. *Let the controller gains be chosen such that theorem 4.3 holds. If the reference r is not constant but has bounded derivatives ($r^{(i)} \in L_\infty[0,+\infty)$ for $i = 0 \ldots n+2$), then the tracking error $\xi_0(t)$ is also bounded.*

Proof. The proof is discussed for the case $n \geq 2$. If r is not constant, then the dynamic system which determines the evolution of $\xi(t)$ is

$$\begin{cases} \dot{\xi} = A(\sigma(t)) + \bar{F}\dot{x}_D - \ddot{x}_D \\ \xi(0) = \dot{e}(0) \end{cases}, \tag{4.71}$$

where $A(\sigma(t))$ is defined in (4.36) and, differently from before, $\bar{F}\dot{x}_D - \ddot{x}_D \neq 0$. Let V be a candidate Lyapunov function,

$$V = \xi^T P \xi, \tag{4.72}$$

whose derivative along the system trajectories is

$$\dot{V} = \xi^T \left(PA(\sigma(t)) + A^T(\sigma(t))P \right) \xi + 2\xi^T P \left(\bar{F}\dot{x}_D - \ddot{x}_D \right). \tag{4.73}$$

Let $Z(\sigma(t)) = PA(\sigma(t)) + A^T(\sigma(t))P$ and $R = P\left(\bar{F}\dot{x}_D - \ddot{x}_D\right)$. Since theorem 4.3 is verified for hypothesis, then $Z(\sigma(t))$ is a symmetric, negative definite matrix for each $\sigma(t)$. The terms $\dot{x}_D, \ddot{x}_D$ are bounded for hypothesis, then the Euclidean norm $\|\cdot\|_2$ of the matrix R is also bounded by a positive constant $\bar{R}$, $\|R\|_2 \leq \bar{R}$. Let $\lambda_{MAX}(\cdot)$ denote the maximum eigenvalue of a matrix. We have that

$$\xi^T Z(\sigma(t)) \xi \leq \lambda_{MAX}\left(Z(\sigma(t))\right) \|\xi\|_2^2. \tag{4.74}$$

If $\bar{\lambda}_{MAX}$ denotes the maximum eigenvalue of $Z(\sigma(t))$ for all $\sigma(t) \in [\sigma_-, \sigma^+]$, i.e.

$$\bar{\lambda}_{MAX} = \max_{\sigma(t)\in[\sigma_-,\sigma^+]} \left\{ \lambda_{MAX}\left(Z(\sigma(t)) \right) \right\}, \tag{4.75}$$

then

$$\xi^T Z(\sigma(t)) \xi \leq \bar{\lambda}_{MAX} \|\xi\|_2^2, \tag{4.76}$$

for all $\sigma(t) \in [\sigma_-, \sigma^+]$. Note that $Z(\sigma(t))$ is negative definite, thus its eigenvalues are negative and real and consequently $\bar{\lambda}_{MAX} < 0$. From (4.73) it is obtained that

$$\dot{V} < \bar{\lambda}_{MAX} \|\xi\|_2^2 + 2\bar{R}\|\xi\|_2 < \|\xi\|_2 \left(\bar{\lambda}_{MAX} \|\xi\|_2 + 2\bar{R} \right), \tag{4.77}$$

which leads to

$$\|\xi\|_2 > -\frac{2\bar{R}}{\bar{\lambda}_{MAX}} \Rightarrow \dot{V} < 0. \tag{4.78}$$

The inequality (4.78) shows that, when the state $\xi(t)$ grows over a certain limit expressed by $-2\bar{R} / \bar{\lambda}_{MAX}$, then V starts decreasing and thus ξ must decrease. Hence, $\xi(t) \in L_\infty[0, +\infty)$ and, obviously, the tracking error ξ_0 is bounded.

The boundedness of the tracking error does not imply the boundness of the integral control action. In fact, if r is not constant, proposition 4.5 does not hold in general. The properties of the integral control action depend on the particular reference r that is chosen. Nevertheless, theorem 4.4 ensures that, if the reference has some temporary changes during operation of the actuator, the error will remain bounded.

Another concern is the robustness of the control algorithm with respect to model uncertainties. The uncertainty about the hysteresis model can be integrated easily in the framework of section 4.2. The imperfect knowledge of the hysteresis relates to an imprecise estimation of the maximum and minimum of $\sigma(t)$. The situation can be described by

$$\sigma_{-} \in [\sigma_{-}^{MIN}, \sigma_{-}^{MAX}], \tag{4.79}$$

$$\sigma^{+} \in [\sigma_{MIN}^{+}, \sigma_{MAX}^{+}], \tag{4.80}$$

where σ_{-}^{MIN}, σ_{-}^{MAX}, σ_{MIN}^{+}, σ_{MAX}^{+}, are known lower and upper bounds of the actual estimations of σ_{-} and σ^{+}.

Proposition 4.5. *If the controller gains are chosen such that theorem 4.3 holds with* $A_{-} = A\left(\sigma(t) = \sigma_{-}^{MIN}\right)$ *and* $A_{+} = A\left(\sigma(t) = \sigma_{MAX}^{+}\right)$, *then the closed-loop (4.35) is asymptotically stable and thus the tracking error converges asymptotically to zero.*

Proof. It comes straightforwardly from the application of the framework.

In section 4.2.2 a performance index of the closed-loop, namely the minimum decay rate α, has been introduced. However, in our case the matrix $A(\sigma(t))$ switches repetitively between A_{-} and A_{+}. The decay α could provide an excessive underestimate of the real decay rate. An *average decay rate* $\alpha_M > 0$ can be defined as the biggest value such that

$$A_M^T P + PA_M + 2\alpha_M P \leq 0 \tag{4.81}$$

holds with $A_M = 0.5A_{-} + 0.5A_{+}$, holds. The matrix A_M is the middle-point of the convex hull determined by A_{-} and A_{+}. As $\sigma(t)$ changes, it is reasonable to assume that the dynamic behavior of (4.35) is, in average, similar to the dynamic behavior established by the eigenvalues of A_M. As will be shown by the experimental results, α_M is a more realistic approximated estimation of the decay rate of the tracking error towards zero, and can be considered a good performance index of the closed-loop.

The average matrix A_M offers also another interesting possibility. The user can exploit the eigenvalues of A_M to design a PID controller with a precise goal, such as a specific transient behavior that, for instance, does not present any overshoot. The design can be carried out with several methods, and a presentation of those possibilities goes beyond the scope of the thesis. They comprehend pole allocation and optimal control (or robust pole allocation and robust optimal control, respectively). On one hand, this offers a clear way to choose the gains of the controller with a performance/goal in mind; on the other hand, it should be remarked that the real performance comes from a PLDI (4.35) and not a linear system, so the resulting choice of the gains has to be intended as an initial choice that could be further tuned if necessary.

At this point, the proposed approach has been discussed in detail. Section 4.2.7 analyses the main differences between this approach and theorems 4.1-4.2.

4.2.7. Comparison with literature

The comparison follows the main issues a-g listed in section 4.1.2.

The proposed approach does not provide the bounds on the controller gains which ensure the asymptotic tracking. Some *candidate* bounds come from the application of the Routh-Hurwitz criterion, but change with respect to the order of the system. See, for instance, the bounds (4.65)-(4.67) for a second order system, or the bounds (4.58)-(4.59) for a first order system. The controller gains must be within the candidate bounds, but the stability and the asymptotic tracking depend on whether the chosen gains satisfy (4.41) or not. An interesting exception is the case of a purely hysteretic system. The bounds (4.50)-(4.51) are necessary and sufficient to guarantee the asymptotic tracking of a constant reference. This confirms the result of theorem 4.2, but extends it. In particular, theorem 4.2 guarantees the existence of a unique solution of the closed-loop and the asymptotic tracking only if the proportional gain is upper bounded by the inverse of the Lipschitz constant, whereas here the proportional gain can be arbitrary positive. It is more complicated to determine a clear difference between the bounds of theorem 4.1 and the candidate bounds proposed here for a second order system, since the latter are only…candidates. In general, they differ. It will be shown by the experimental results that in some cases the proposed method offers larger bounds of the controller gains than theorem 4.1. However, the proposed method requires the condition (4.41) to be checked, and this could require some further numerical effort.

Theorems 4.1 and 4.2 do not determine the type of convergence of the tracking error around zero, but only state that the error decreases asymptotically. The proposed method, instead, ensures an exponential convergence, at a decay rate that is at least equal to the minimum decay rate α. This rate, however, overestimates the time of convergence, and the average decay α_M in (4.81) is a better estimation. Again, an interesting result is given by the purely hysteretic system considered in theorem 4.2. It proves that the tracking error decreases monotonically without overshoots. Refer to the closed-loop (4.48). It turns out to be a first order PLDI, where for any trajectory of the tracking error there is just one eigenvalue. In other words, it is very difficult that an underdamped behavior typical of second-order systems arises, thus confirming the result of theorem 4.2 but extending it to any positive value of the gains.

It has been shown in theorem 4.4 that the proposed method guarantees a bounded tracking error in the case of a non-constant reference. This is a particular feature with respect to theorem 4.1, which does not discuss the case. Theorem 4.2, indeed, proves the BIBO stability of the closed-loop with unity gain in the case of an arbitrary reference.

A critical point of theorem 4.1 and 4.2 is that no way is given to choose the controller gains. Only bounds are provided. Then, the user should tune the PID controller based on his personal knowledge on the system, or using some simulation software to test several gains and find the best ones for his application. On the other hand, the proposed method offers a way to search for the gains. This way is approximate, since it relies on the exploitation of the average dynamics, which do not entail a knowledge of the actual system, which is a PLDI. Nevertheless, the user can design a controller with well-known methods, such as pole allocation or optimal control, and then test if the desired performances are obtained experimentally. At least, there exists a starting point of the gains, which is directly based on a clear desired behavior or performance. Further research can focus on a new definition of the average dynamics in order to make it more representative of the PLDI, and more useful for controller design.

Another critical point is that the approach of theorems 4.1 and 4.2 is particular for the case that they discuss, i.e., only for second order systems and purely hysteretic ones. The paper [46] addresses only pure integral control of high order systems. The proposed approach, as shown in sections 4.2.2-4.2.5, can deal with high order systems as well as low order ones, assumed that they can be described by (4.1)-(4.3).

A delicate point of the proposed approach concerns the handling of model uncertainties. If the hysteresis is not known with precision, then proposition 4.5 can be used to solve the problem. If the uncertainty regards the dynamic parameters, i.e. the matrix F in (4.2), things become more complicated and the convex hull of $A(\sigma(t))$ does not have only two vertices (A_{-}, A_{+}) anymore. Methods to address this case are described in [51]. A critical point of the proposed approach concerns indeed the handling of a time-varying hysteresis. Theorem 4.1, in fact, can be still used if the hysteresis changes with time, since this change relates to a Lipschitz constant that is a function $L_C = L_C(t)$. The proposed approach cannot deal with time-varying nonlinearities in such an easy way. This can be the focus of future research.

Unfortunately, the proposed method shares with theorems 4.1 and 4.2 an important limitation: the conditions for stability are only sufficient, and consequently the bounds on the gains are conservative. In fact, the method makes use of the *quadratic stability* concept for the PLDI (4.35), which is known to be a conservative one for PLDIs. However, an important result [51] states that a *PLDI is exponentially stable if and only if it admits a polyhedral Lyapunov function.* Thus, the proposed approach can lead to less conservative bounds on the gains, and less conservative control laws. This can also be an issue for future research.

4.3. Experimental results

The method proposed in section 4.2 has been applied to two MSM actuators. The present section summarizes the results.

4.3.1. First actuator

The results of this section refer to the actuator in Figure 2.5, and come out of a collaboration with TU Braunschweig in Germany [48]. In order to proceed with the stability analysis, the actuator system has been characterized in terms of the series between a hysteresis operator and linear dynamics, as in Figure 4.1.

The chosen hysteresis model is the MPIM. The comparison between the actual hysteresis and the identified one (dashed) is reported in Figure 4.3. The identified model has $\sigma_{-} = 0.44\ \mu\text{m/A}$ and $\sigma^{+} = 420.25\ \mu\text{m/A}$. The dynamic part has been identified with a first order linear system as specified in (4.52), with $\chi = \beta = 99.2\ \text{1/s}$.

We discuss here two PI controllers, PI1 and PI2 respectively.

The controller PI1 is characterized by gains $K_P = 0.00001\ \text{A/}\mu\text{m}$, $K_I = 0.01\ \text{A/(s}\cdot\mu\text{m)}$. A simple numerical check of (4.41) states that those gains ensure the stability of the closed-loop (described by the matrix (4.53)). The average decay rate is $\alpha_{M1} = 1.114\ \text{1/s}$, and the corresponding time-constant is $\tau_{M1} = 1/\alpha_{M1} = 0.89\ \text{s}$. The settling-time can be estimated as $4\tau_{M1} = 3.5\ \text{s}$. To analyze the (average) transient behavior of the closed-loop, the matrix A_M should be characterized in terms of its eigenvalues. The eigenvalues of the matrix A_M are the roots of the equation

$$s^2 + (\chi + \beta K_P \bar{\sigma}) s + \beta K_I \bar{\sigma} = 0 , \tag{4.82}$$

where $\bar{\sigma} = (\sigma_- + \sigma^+)/2 = 210.34$ µm/A . Equation (4.82) describes a second-order system and can be rewritten in terms of damping ratio δ and natural frequency ω_N . PI1 leads to the value $\delta > 1$, so that the average closed-loop dynamic exhibits an overdamped behavior without overshoot. There is a pole at -2.15 rad/s , and a pole at -97.2 rad/s .

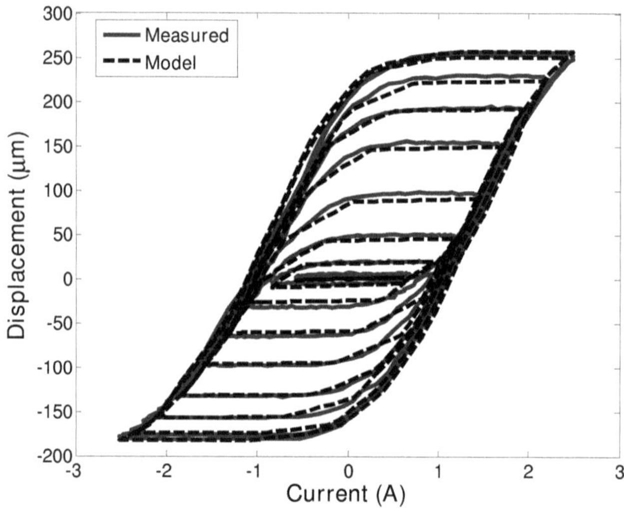

Figure 4.3 – Identification of the actuator hysteresis by a MPIM

Figure 4.4 shows the estimated average decaying exponential of the tracking error given by PI1. Figure 4.5 (top) shows the result of the experimental application of PI1. The asymptotic tracking is ensured; however, the closed-loop is quite slow. Note that there is no overshoot, as predicted by the average dynamics which are overdamped. The zoom of Figure 4.5 (bottom) shows that the predicted settling-time of $4\tau_{M1}$ is quite accurate. It can be noted that the predicted time-constant $\tau_{M1} = 0.89$ s differs from the time-constant associated to the eigenvalues of A_M , that is $\tau_1 = 0.46$ s. This is due to the fact that τ_{M1} is strictly connected to the matrix P which depends on the gains; consequently, it entails some knowledge of the PLDI (4.35), whereas τ_1 does not.

The second design example shows how an initial choice of the controller gains can be done based on some desired requirements. In fact, (4.82) can be used in the other direction: given a desired average behavior of the closed-loop system, for instance in terms of desired values δ^* and ω_N^*, the gains K_P and K_I of the controller PI2 can be determined. Let us choose the average behavior characterized by $\delta^* = 0.628$ and $\omega_N^* = 79.1$ rad/s, in order to have a system which is faster than the one with PI1 and has a small overshoot. From (4.82) it comes that the desired gains of the controller PI2 are $K_P = 0.00001$ A/μm and $K_I = 0.3$ A/(s · μm). With these gains, condition (4.41) must be checked to conclude about stability. In this case it is satisfied. The resulting average decay rate is $\alpha_{M2} = 19.076$ 1/s, which leads to a time-constant $\tau_{M2} = 0.0524$ s and a settling-time of $4\tau_{M2} = 0.21$ s.

Figure 4.4 shows the predicted decaying exponential related to the closed-loop with PI2, which is more than ten times faster with respect to PI1. Figure 4.6 reports the experimental result using PI2. The zoom at the bottom allows appreciating that the predicted settling-time is quite accurate. Moreover, an overshoot can be seen. The predicted average value of this overshoot, based on δ^* and ω_N^*, is 8%. The real overshoot is a little bit smaller than the estimated (average) one, as can be seen in Figure 4.6. However, once again the settling-time and the overshoot predictions show to be good estimations of the behavior of the closed-loop. As before, it can be noted that the time-constant $\tau_{M2} = 0.0524$ s is bigger than the time-constant associated to the roots of (4.82), which is $\tau_2 = 0.023$ s. Decreasing τ_{M2} means increasing the controller gains (especially K_I), and this cannot be always done. In the case that the performance has to be improved, then the designer could go back to the definition of the target average behavior and carry out another design.

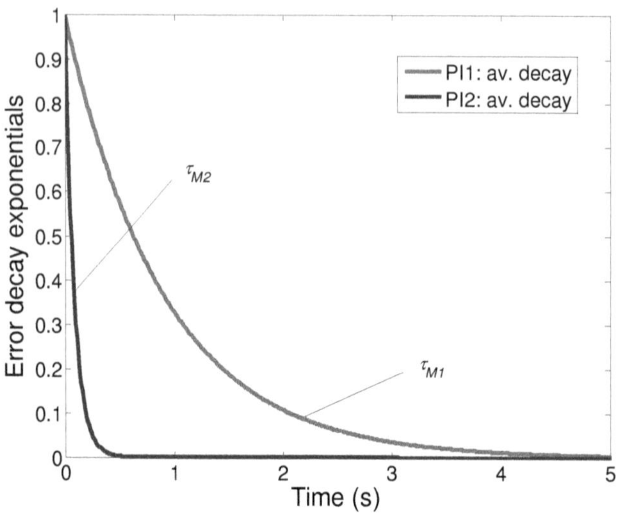

Figure 4.4 – Estimated (average) decay times for the closed-loops with PI1 and PI2

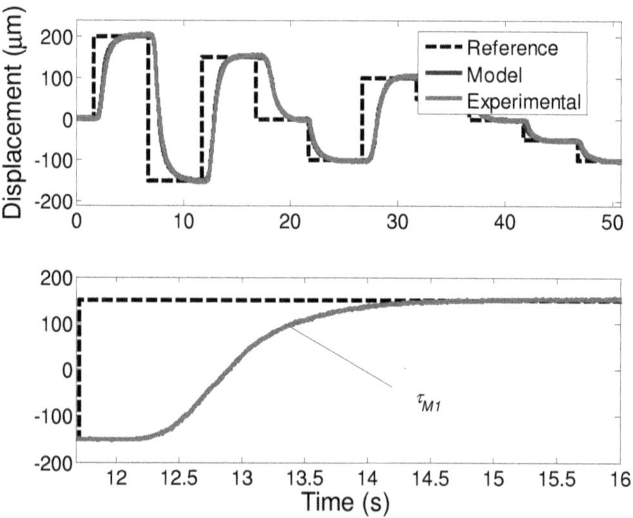

Figure 4.5 – Tracking with PI1 (top) and zoom on one step (bottom)

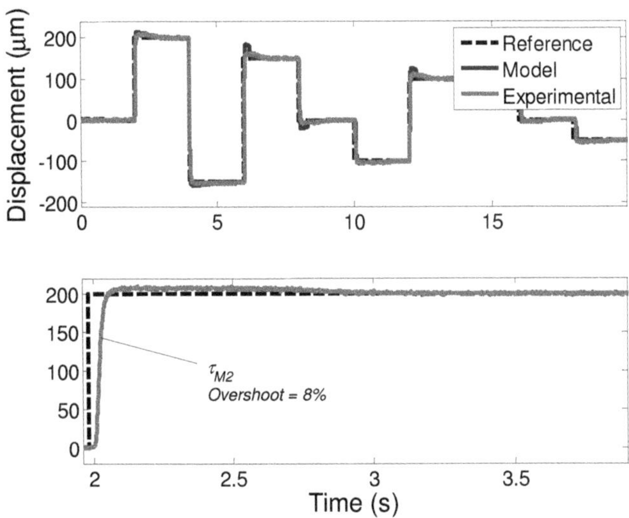

Figure 4.6 – Tracking with PI2 (top) and zoom on one step (bottom)

4.3.2. Second actuator

The results of this section refer to the actuator in Figure 2.4, and come out of a collaboration with the Finnish enterprise Adaptamat. The actuator system has been characterized again in terms of the series between a hysteresis operator and a linear dynamics, as in Figure 4.1. The results reported here have been illustrated in [49].

The chosen hysteresis model is the MPIM. The comparison between the actual hysteresis and the identified one (dashed) is reported in Figure 4.7. The identified model has $\sigma_{-} = 3.7$ μm/A and $\sigma^{+} = 620.7$ μm/A. The dynamic part has been identified with a second order linear system as follows

$$\Sigma : \begin{cases} \dot{x}_1 = x_2 \\ \dot{x}_2 = -2\delta\omega_N x_2 - \omega_N^2 x_1 + \omega_N^2 \Gamma[u] \\ y = x_1 \end{cases}, \tag{4.83}$$

where natural frequency ω_N and damping coefficient δ are used instead of *m, k, c*. In particular, $\delta = 0.63$ and $\omega_N = 1000$ rad/s .

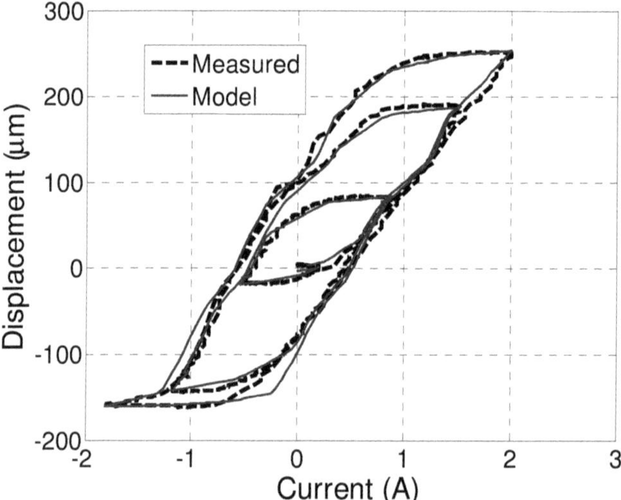

Figure 4.7 – Identification of the actuator hysteresis by a MPIM

We discuss here three different PI controllers. The derivative gain has been kept equal to zero because of the presence of measurement noise in the control-loop.

If $K_P > 0$, and considering the formulation of the dynamics in terms of δ and ω_N , the only bound is on the integral gain, and from (4.66) it becomes

$$K_I < \frac{\left(2\delta\omega_N + K_D\omega_N^2\sigma_-\right)\left(1 + K_P\sigma_-\right)}{\sigma^+}. \tag{4.84}$$

As before, in the following pictures the tracking results are compared also with the tracking behavior predicted by the simulation model of the MSM actuator, composed by the series of the identified MPIM and the identified dynamics.

Figure 4.8, Figure 4.9 and Figure 4.10 show the effects of increasing the integral gain while the proportional gain is kept constant, $K_P = 0.001$ A/μm. The settling time of the closed-loop is reduced from 1.15 s of Figure 4.8, to 0.7 s of Figure 4.9 and then to 0.1 s of Figure 4.10. In each figure (bottom part) the tracking error versus time is shown (expanded view of one step), and its decaying behavior is compared with the estimated decaying exponentials related to the average decay rates α_M (dashed lines). It is possible to appreciate what already shown and stated: the exponential $e^{-\alpha_M t}$ gives a good prediction of the actual error behavior, whereas the exponential related to the minimum decay rate would give an excessive overestimation of the decaying time. It can be concluded that α_M represents a good performance index of the obtainable closed-loop performance. Figure 4.11 shows the effect of the application of a sinusoidal reference signal. As stated by theorem 4.4, the error is bounded; however, the tracking performance cannot be guaranteed as for constant references. It is worth noting that the model prediction is quite good in every case, thus confirming the usefulness of the modeling and identification approaches mentioned in section 2.3.1.

The application of the control design proposed in [41] for PI controllers, with the Lipschitz constant of the identified hysteresis $L_C = 1200$ μm/A, leads to the bounds $K_P < c^2 / L_C m = 0.0007$ A/μm and $K_I < K_P k / c = 0.5764$ A/(s · μm). In the presented experiments and simulations, the values $K_P = 0.001$ A/μm and $K_I = 0.208$ A/(s · μm), $K_P = 0.002$ A/μm and $K_I = 0.61$ A/(s · μm) have been used, and for the approach presented here they ensure the asymptotic tracking of constant references. This points out that the presented approach could lead to larger parameters ranges than the one in [41] (at the cost of further numerical efforts).

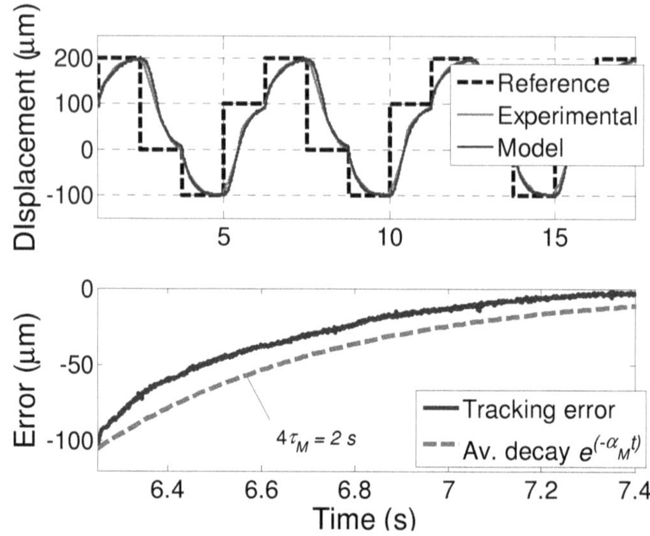

Figure 4.8 – Tracking with *Ki = 0.02* (top) and tracking error (bottom)

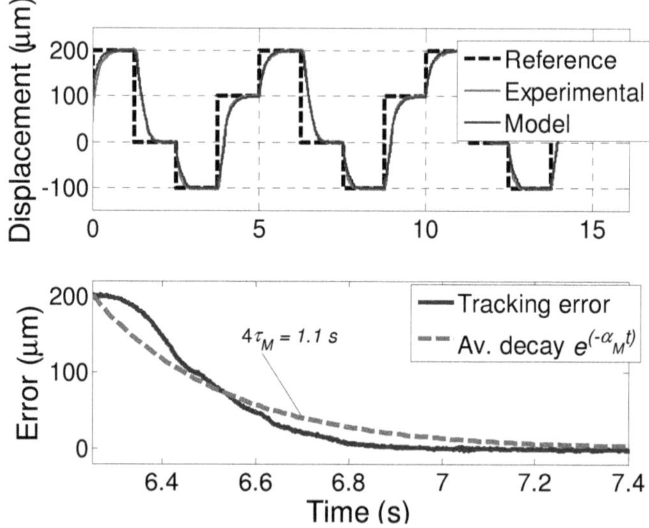

Figure 4.9 – Tracking with *Ki = 0.041* (top) and tracking error (bottom)

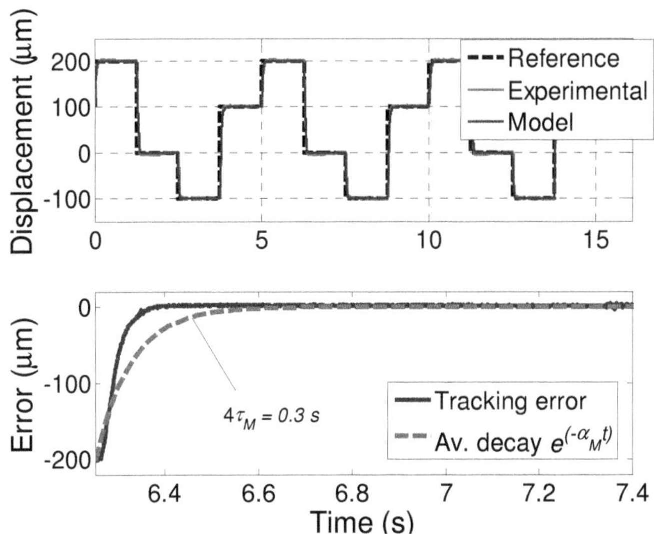

Figure 4.10 – Tracking with *Ki = 0.208* (top) and tracking error (bottom)

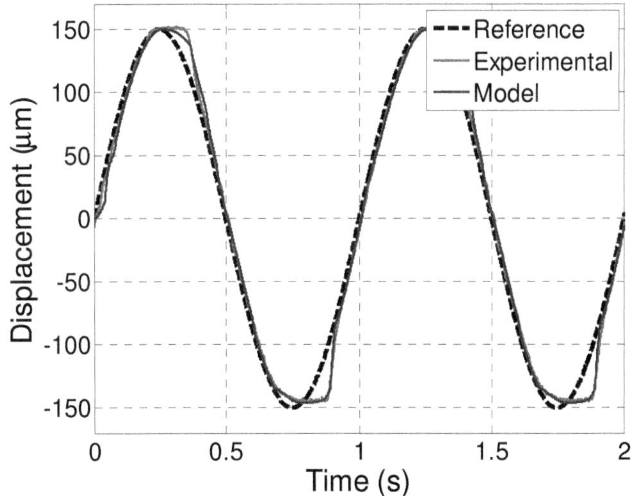

Figure 4.11 – Tracking of a non-constant reference (*Ki = 0.61, Kp=0.002*)

5. Adaptive control

Figure 5.1 shows a representation of the closed-loop for the control of hysteretic dynamic systems. The reference x_D is now not required to be constant as in the previous chapter. The control objective of the controller C is to ensure that the output variable y follows the reference trajectory x_D. Hysteresis strongly influences the tracking result, as discussed for instance in [52]. Literature offers a wide range of solutions for the tracking control of hysteretic dynamic systems. Basically, there are three main approaches.

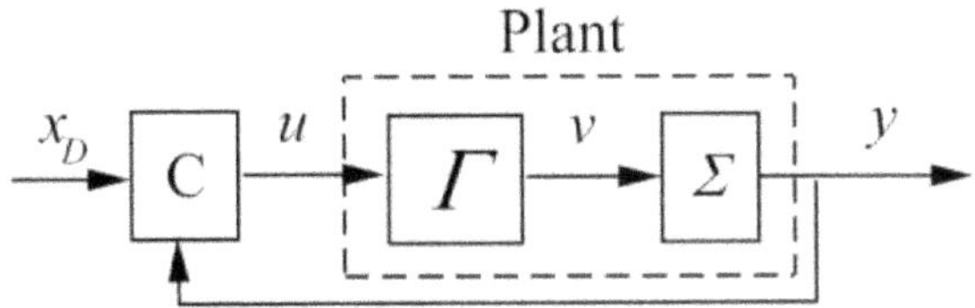

Figure 5.1 – Control-loop for tracking purposes

Open-loop compensation requires a mathematical model of the hysteresis to be built and inverted on-line to achieve a linearization of the input-output $x_D - y$ characteristic. The controller C of Figure 5.1 is composed only by the hysteresis compensator, and there is no direct feedback from y. This strategy can rely on the maturity of several models such as the ones presented in chapter 3. Also *physics-based models* [8], which incorporate the basic knowledge about the physics of the system under analysis, have demonstrated to be useful for open-loop compensation purposes. The open-loop compensation approach has been shown to be particularly effective when the actuator operates in frequency ranges in which dynamic phenomena can be neglected.

Closed-loop control approaches do not explicitly develop a model for the hysteresis, but rather attempt to compensate for its effects by proper tuning of the control law (e.g. [41], [33], [53]). For instance, in [54] the authors analyze the *dissipation* properties of hysteresis, and in [55] they exploit those properties to achieve robust control algorithms based on passivity theorems. The PID control discussed in chapter 4 belongs to this class of approaches.

Hybrid strategies bring together the first and second type of approaches. The use of a feedforward compensator can lead to an easier closed-loop control design, for instance making it possible to assume that the residual dynamics of the compensated plant are linear. Moreover, it sometimes improves the requirements on the control action by reducing its amplitude and improving the overall behavior [56], [57]. In [58] the authors make use of a Preisach compensator together with a PI controller to improve the tracking performance of a piezoelectric actuator. In [59] a robust sliding mode controller is proposed to counteract the effect of disturbances and model errors on the compensation path; in [60] a lead-lag controller is used to handle the remaining dynamics of the plant. In [61] the adoption of a Preisach-like compensator is used in conjunction with a disturbance observer to compensate for residual errors. The authors of [62] and [63], instead, propose a robust and adaptive control strategy that exploits the compensation abilities of a physics-based model of magnetostrictive materials. Reference [34] investigates experimentally the efficacy of a standard PI controller together with a MPIM compensator for position control with MSM actuators. In [64], the displacement produced by a MSM actuator is controlled by a controller which entails an adaptive MPIM compensator, while the residual compensation errors are addressed by adaptive bounding strategies.

All the mentioned references assume either that a model of the hysteresis or some characteristics of the nonlinearity are known with some accuracy, either that an accurate compensator is available. A challenging situation is represented by those systems which exhibit a time-varying hysteresis, changing in shape and offset due to stress [65], temperature [66] or other influencing variables. MSM actuators belong to this class (see Figure 2.10 and Figure 2.11), as documented in [57]. In this case, in fact, an open-loop compensator that has been tuned on a specific input-output curve will fail to achieve a satisfactory performance. On the other hand, closed-loop and hybrid strategies can also be negatively affected by time-varying hysteretic phenomena, because the control parameters are generally optimized for a particular hysteresis curve, or rely on compensation by accurate inverse models. For example, in the case of variations of the hysteresis characteristic, the strategy proposed in [59] will see the expected compensation error growing, and possibly going over the bound that has been used for the design of the robust control contribution, making the provided stability analysis not valid anymore. The analysis provided in [64], instead, will still hold since the robust term is adaptive, but the magnitude of this term will increase and probably bring the control action into saturation, with a negative consequence over the tracking performances.

A possible solution is the use of adaptive hysteresis compensators that identify and invert the hysteresis curve on-line, thus ensuring small compensation errors also in the case of a variation of the characteristic. This approach has been introduced by Tao and Kokotovic in [67], where they define and adapt a simple *local* hysteresis model. Recent examples of this approach are the adaptive gradient-based version of the KPM proposed in [39], and the adaptive PIM in [68] which uses gradient-based adaptation to deal with the hysteresis and creep effects of a piezoelectric actuator. Reference [69] addresses the development of an adaptive MPIM for the compensation of the hysteresis in a piezo-based positioning actuator. Unfortunately, references [39], [68], [69] do not consider the presence of dynamics in the plant, and thus they belong to the class of *adaptive open-loop compensation approaches*. They can be used only when the plant dynamics are neglectable.

References [70], [71] propose a robust adaptive control algorithm for nonlinear dynamic systems with an input hysteresis described by a generalized PIM. The nonlinear functions of the dynamics are assumed to be known. The control strategy does not entail a compensator, but estimates the contribution of the hysteresis and uses it within a sliding mode control framework. A robust term that switches sign depending on the tracking error, and whose amplitude is related to the hysteresis contribution, appears in the control law.

Reference [72] investigates the effectiveness of a control-loop where a KPM compensator (refer to section 3.2.4) is used to linearize the system between u and v, and a further linear model reference adaptive controller [73] is used to shape the remaining dynamics Σ, assumed linear. The adaptive laws of the compensator and the controller are derived with the MIT rule [73]. The paper shows that, if the adaptive gains are chosen in a proper way, then the hysteresis model parameters adapt towards the optimal values and the tracking of a desired reference is ensured. A key assumption of the paper is that the hysteresis phenomenon changes very slowly, but this is not always true in the case of MSM-based actuators.

This chapter proposes another approach. We consider that the dynamics of the plant are nonlinear and unknown. The hysteresis, which is time-varying, is described by a MPIM. The controller C entails a MPIM compensator (IMPIM) that is adaptive, so that its parameters change based on the tracking error, with mathematical laws that are defined later on. In particular, the proposed framework allows adapting also the thresholds of the SO, while in related literature of the MPIM these thresholds are usually fixed a priori. The control law, which is the input to the adaptive compensator, is based on a feedback linearization strategy: the unknown nonlinear dynamics of the plant are estimated on-line and then canceled by the control action. The asymptotic convergence of the tracking error $y - x_D$ to zero is demonstrated based on Lyapunov arguments. There is no particular assumption about the rate at which the hysteretic phenomenon changes.

Section 5.1 describes in more details the tracking problem and illustrates the representation of the plant that is assumed for the developments of the other sections.

Section 5.2 recalls some known results about function approximation: the unknown nonlinear dynamics of the plant are in fact approximated by suitable function approximators, which are introduced. The mathematical notation used thorough this chapter is also introduced in section 5.2. Particular emphasis is given to the hysteresis. In section 5.2.4 it is demonstrated that the SO is a universal function approximator.

Section 5.3 develops the mathematical framework for the adaptive control of the plant. The main issues are discussed in details in sections 5.3.1-5.3.3. Section 5.3.4 summarizes and orders the main steps of the control algorithm.

Section 5.4 presents a simplification of the adaptive control approach of section 5.3 for plants which are characterized by first order linear dynamics. This allows focusing the discussion on the adaptive hysteresis compensator, and analyzing it in two simulation scenarios which show the importance of the thresholds adaptation. Another reason for the simplification is that in the experimental results the plant has been assumed as a first-order linear system.

Section 5.5 presents the experiments performed on the MSM positioning actuator of Figure 2.4. The effectiveness of the adaptive control approach is validated in several scenarios with different reference signals. The last experiment has been done to evaluate the approach in the challenging situation where the actuator is subjected to temperature disturbances. It is demonstrated that the proposed control strategy ensures a constant tracking performance also in the case of bounded temperature disturbances.

5.1. Statement of the control problem

The control-loop that is considered in this chapter is sketched in Figure 5.2.

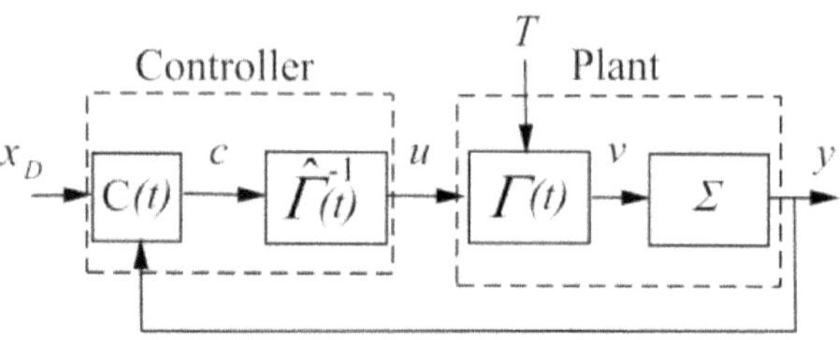

Figure 5.2 – Control-loop for adaptive control

The hysteretic plant is composed by the series of a time-varying hysteresis operator $\Gamma(t)$ and a time-invariant nonlinear dynamic Σ that is assumed to be described by

$$\Sigma:\begin{cases}\dot{x}_1 = x_2\\ \dot{x}_2 = x_3\\ \vdots\\ \dot{x}_n = f_0(x)+f(x)+(g_0+g)v\\ y = x_1\\ x(t_0)=x(0)\end{cases}, \tag{5.1}$$

where $v\in\Re$ is the input signal (the output of the hysteresis operator); $f_0(x): X\subset\Re^n\to\Re$ is the known part of the nonlinear dynamics while $f(x): X\subset\Re^n\to\Re$ is the unknown part; $g_0\in\Re$ is the known input gain, while $g\in\Re$ is the unknown one. The state vector $x=[x_1,x_2,...,x_n]^T$ belongs to the compact set $X\subset\Re^n$, and the nonlinear functions are continuously differentiable in X, $f_0, f\in C^1(X)$. The dynamic system Σ can also be expressed by

$$\dot{x} = Ax + B\left(f_0(x) + f(x) + \left(g_0 + g\right)v\right), \quad x(t_0) = x(0), \tag{5.2}$$

where

$$A = \begin{bmatrix} 0 & 1 & 0 & \dots & 0 \\ 0 & 0 & 1 & \dots & 0 \\ \vdots & \vdots & \vdots & \ddots & \vdots \\ 0 & 0 & 0 & \dots & 1 \\ 0 & 0 & 0 & \dots & 0 \end{bmatrix} \in \Re^{n \times n}, \quad B = \begin{bmatrix} 0 \\ 0 \\ \vdots \\ 0 \\ 1 \end{bmatrix} \in \Re^{n}. \tag{5.3}$$

The variation of the hysteresis depends on the influence of temperature, which is *not measured* during operation. The input to the hysteresis operator is the *actual control variable u*, which is generated by the application of an estimated adaptive inverse hysteresis model $\hat{\Gamma}^{-1}(t)$ to the *control variable c*. If the compensation is perfect then $v \equiv c$. The control variable c is the output of the adaptive controller $C(t)$, which elaborates the tracking error $\tilde{x}$, i.e. the difference between the measured trajectory x and the desired trajectory x_D

$$\tilde{x} = x - x_D, \tag{5.4}$$

where x_D contains the desired displacement x_{D1} and its derivatives $x_{D1}^{(i)}$ for $i = 1..n-1$

$$x_D = [x_{D1}, \dots, x_{D1}^{(n-1)}]^T \in \Re^n. \tag{5.5}$$

The derivatives of the desired displacement are assumed to be available and bounded up to the n-th order.

The control objective is to make the measured variable (the displacement in the case of MSM actuators) $y = x_1$ follow the desired trajectory x_{D1}, when the hysteresis is *almost unknown* and the dynamics of the plant are only *partially known* or even *completely unknown*. Temperature is assumed not to be measured. References [74], [75] report two examples of model-scheduling compensation strategies which make use of temperature measures to control the displacement of MSM actuators.

5.2. Function approximation

In this brief section the basic concepts of function approximation are recalled. In particular, the emphasis is on piecewise linear approximators, which will be used to identify and compensate for the unknown functions that influence the nonlinear dynamics (5.1).

5.2.1. Basic concepts

Let $f : X \subset \Re^n \to \Re$ be a nonlinear real valued function defined in the compact set X and continuously differentiable in X, $f \in C^1(X)$. The following definitions introduce the concept of *local approximation* and *global approximation*.

Definition 5.1 [76], [77]. *A function* $\hat{f}(x;\hat{\theta})$ *is a local approximation to* $f(x)$ *at* x_0 *if for any* ε *there exist* $\hat{\theta}$ *and* δ *such that*

$$\left\| f(x) - \hat{f}(x;\hat{\theta}) \right\| \leq \varepsilon \text{ for each } x \in I(x,\delta), \tag{5.6}$$

where $I(x_0,\delta) = \left\{ x \in X \mid \left\| x - x_0 \right\| \leq \delta \right\}$.

Definition 5.2 [76], [77]. *A function* $\hat{f}(x;\hat{\theta})$ *is an* ε*-accurate approximation to* $f(x)$ *over the domain* X *if for the given* ε *there exists a* $\hat{\theta}$ *such that*

$$\left\| f(x) - \hat{f}(x;\hat{\theta}) \right\| \leq \varepsilon \text{ for all } x \in X. \tag{5.7}$$

Definition 5.2 characterizes the property of a global approximation to a certain unknown function $f(x)$. The global approximation can also rely on many local approximations which are composed in a convenient way, as stated in the following definition.

Definition 5.3 [77]. *A function approximator* $\hat{f}(x;\hat{\theta})$ *is of the Basis-Influence (BI) class if and only if it can be written as*

$$\hat{f}(x;\hat{\theta}) = \sum_{i}^{N_f} f_i(x;\hat{\theta})\mu_i(x), \tag{5.8}$$

where each $f_i(x;\hat{\theta})$ *is a local approximation to* $f(x)$ *for all* $x \in I(x_i,\delta)$, *and* $\mu_i(x)$ *has local support* $S_i \subset I(x_i,\delta)$ *such that* $X \subseteq \bigcup_i S_i$.

In definition 5.3, the overall approximation structure $\hat{f}(x;\hat{\theta})$ is the composition of N_f local approximation structures which are "centered" on the N_f nodes $x_i \in X$. The *influence functions* $\mu_i(x)$ determine to which extent the local approximators are local.

Definition 5.4 [77]. *The set of positive semidefinite influence functions* $\{\mu_i(\cdot)\}$ *form a partition of unity on* X *if for any* $x \in X$,

$$\sum_{i=1}^{N_f} \mu_i(x) = 1. \tag{5.9}$$

It can be noted that, if the property (5.9) is satisfied, then the overall approximator $\hat{f}(x;\hat{\theta})$ is the convex combination of $f_i(x;\hat{\theta})$ for each $x \in X$. Property (5.9) is fundamental for the following result.

Theorem 5.1 [77]. *If* $\hat{f}(x;\hat{\theta})$ *belongs to the BI class with each* $f_i(x;\hat{\theta})$ *being a local* ε*-accurate approximation of* $f(x)$ *(see definition 5.1) and* $\mu_i(x) \geq 0$, *then*

(5.9) is a sufficient condition for $\hat{f}(x;\hat{\theta})$ to be an ε-accurate global approximation of any $f \in C^1(X)$.

5.2.2. Piecewise linear approximators

A piecewise linear approximator $\hat{f}(x;\hat{\theta})$ interpolates the unknown function with a set of local approximators which are linear. In mathematical terms,

$$\hat{f}(x;\hat{\theta}) = \sum_{i=1}^{N_f} \left(\alpha_i^T (x-\zeta_i) + \beta_i \right) \mu_i(x), \tag{5.10}$$

where $\zeta_i \in \Re^n$ indicates the center of the i-th local region, $\alpha_i \in \Re^n$ indicates the slope and $\beta_i \in \Re$ is the known coefficient [78]. Typical choices of the influence functions are

$$\mu_i(x) = \begin{cases} 1 & \text{if } \|x-\zeta_i\| \le \delta \\ 0 & \text{otherwise} \end{cases}, \tag{5.11}$$

$$\mu_i^0(x) = \begin{cases} e^{-\|x-\zeta_i\|} & \text{if } e^{-\|x-\zeta_i\|} \ge \delta \\ 0 & \text{otherwise} \end{cases}, \tag{5.12}$$

where δ is a design parameter that determines the extension of the local support S_i of the basis functions $\hat{f}_i(x;\hat{\theta}) = (\alpha_i^T(x-\zeta_i)+\beta_i)$. For influence functions like (5.11), the validity of the property (5.9) comes straightforwardly from a simple choice of δ. For functions like (5.12), the locally defined $\mu_i(x)$ which satisfy (5.9) can be obtained as follows [77]

$$\mu_i(x) = \frac{\mu_i^0(x)}{\sum_{i=1}^{N_f} \mu_i^0(x)} . \tag{5.13}$$

Thanks to theorem 5.1, it is possible to state that a piecewise linear approximator (5.10) with basis functions $\hat{f}_i(x;\hat{\theta}) = (\alpha_i^T(x-\zeta_i)+\beta_i)$ and influence functions defined by (5.11) or (5.13)-(5.12), is a ε-accurate global approximator of any $f \in C^1(X)$. The accuracy of the approximator depends on the number N_f of basis functions. Equation (5.10) can be manipulated in the standard form

$$\hat{f}(x;\theta) = \theta^T \phi_f(x) , \tag{5.14}$$

where

$$\theta = [\alpha_1^T, \beta_1, ..., \alpha_{N_f}^T, \beta_{N_f}]^T , \tag{5.15}$$

$$\phi_f(x) = [(x-\zeta_1)^T \mu_1(x), \mu_1(x), ..., (x-\zeta_{N_f})^T \mu_{N_f}(x), \mu_{N_f}(x)]^T . \tag{5.16}$$

5.2.3. Notation

Let $f: X \subset \Re^n \to \Re$ be a nonlinear real valued function defined in the compact set X and continuously differentiable in X that has to be approximated. In the following the approximator of $f(x)$ is indicated by $\hat{f}(x) = \hat{f}(x;\hat{\theta})$, where $\hat{\theta}$ represents the actual estimate of the parameter (*weight*) vector θ (defined in (5.15) for a piecewise linear approximator). The fact that the overall approximator (5.10) has a ε-accuracy over the approximation region X, points out that the approximation error within X can be made as small as desired by a proper choice of N_f. However, it will never be zero. Within the family of approximators (5.14), we denote with $f^*(x) = \hat{f}(x;\theta^*)$ the best approximator, i.e. the approximator that has the best approximation accuracy over X. The best approximator is determined by the optimal weight vector θ^*, which can be formally defined by

$$\theta^* = \arg\min_{\hat{\theta}\in\Re^{N_f}} \left\{ \sup_{x\in X} \left| f(x) - \hat{f}(x;\hat{\theta}) \right| \right\}. \tag{5.17}$$

The best approximator $f^*(x)$ satisfies the following relationship with $f(x)$:

$$f(x) = f^*(x) + \varepsilon_f(x;N_f), \tag{5.18}$$

where $\varepsilon_f(x;N_f)$ is called the *minimum functional approximation error* (MFAE), and is the unavoidable deviation between the function to be approximated and the best approximation. Obviously, $\varepsilon_f(x;N_f)$ depends on the particular family of approximators, and on the number N_f of basis functions. The *parameter deviation* or *parameter error* $\tilde{\theta}$ is defined as the difference between the actual estimation and the best value,

$$\tilde{\theta} = \hat{\theta} - \theta^*. \tag{5.19}$$

5.2.4. Approximation of the hysteresis

A very similar discussion can be done concerning the approximation of the hysteresis phenomenon by means of the hysteresis models introduced in chapter 3. Let Γ be a hysteresis operator. Theorem 5.1 is not valid anymore, since Γ is not a memoryless function, and thus it cannot be approximated globally by a common piecewise linear approximator. It has been shown that the MPIM and the KPM are indeed adequate hysteresis approximators. The symbol $\hat{\Gamma}$ denotes the approximation of the actual hysteresis phenomenon Γ, which depends on the actual parameters estimates. In the case of a KPM,

$$\hat{\Gamma}[u] = \hat{\Gamma}[u; \hat{w}, \hat{a}, \hat{\alpha}, \hat{\beta}], \tag{5.20}$$

and in the case of a MPIM,

$$\hat{\Gamma}[u] = \hat{\Gamma}[u; \hat{w}, \hat{p}, \hat{r}, \hat{a}], \tag{5.21}$$

where the meaning of each parameter can be recalled from chapter 3. In the following the main focus will be on the MPIM, although most of the mathematical developments can be transferred straightforwardly to the KPM. The best approximation

$$\Gamma^*[u] = \hat{\Gamma}[u; w^*, p^*, r^*, a^*] \tag{5.22}$$

is given by the optimal parameters w^*, p^*, r^*, a^*, such that

$$\Gamma[u] = \Gamma^*[u] + \varepsilon_\Gamma(u; N, L), \tag{5.23}$$

where $\varepsilon_\Gamma(u; N, L)$ is the MFAE of the hysteresis and N, L are the orders of the MPIM. The MFAE depends, as before, on the chosen model family (MPIM) and on the orders of the model. The parameters deviations are defined in a similar way as before, i.e.,

$$\tilde{w} = \hat{w} - w^*, \tag{5.24}$$

$$\tilde{p} = \hat{p} - p^*, \tag{5.25}$$

$$\tilde{r} = \hat{r} - r^*, \tag{5.26}$$

$$\tilde{a} = \hat{a} - a^*. \tag{5.27}$$

The best hysteresis approximation relates to the best compensator Γ^{*-1}, which ensures the best inversion of the hysteresis phenomenon. We recall that the compensator $\hat{\Gamma}^{-1}$ of the MPIM can be calculated analytically from the actual estimates $\hat{w}, \hat{p}, \hat{r}, \hat{a}$ of the direct model $\hat{\Gamma}$.

Quite often, some model parameters are not adapted on-line but instead fixed a priori, based on the preliminary knowledge of the phenomenon. For instance, in the identification of the KPM and the MPIM, the thresholds are fixed and the identification procedure searches only the optimal values of the weights. Obviously, this choice influences the value of the MFAE ε_{Γ}.

The hysteretic phenomenon of MSM actuators presents a time-varying offset, where the offset is calculated with respect to the initial (0,0) point of the current-displacement characteristic. Refer to Figure 2.10 or Figure 2.11: when temperature increases, the hysteresis changes its shape and offset. The hysteresis operator Γ is then supposed to be able to capture the presence of an offset in the $u-v$ characteristic. Due to the nature of the play operators defined in (3.21), the MPIM model in (3.44) cannot entail the presence of an offset, and thus the offset must be added externally. With this in mind, in the following the hysteresis to be approximated is indicated by $\Gamma + \sigma$, where σ is the time-varying offset. The approximation is indicated with $\hat{\Gamma} + \hat{\sigma}$, where $\hat{\Gamma}$ is defined in (5.21) and $\hat{\sigma}$ is the actual estimation of the offset σ. Consequently, the best approximation is indicated with $\Gamma^* + \sigma$, where Γ^* is defined in (5.22). The offset deviation is

$$\tilde{\sigma} = \hat{\sigma} - \sigma . \tag{5.28}$$

The MPIM is the series of a hysteresis model (the PIM) and a piecewise linear function, the SO, that can be proved to be a universal function approximator of the BI class. Let the input to the SO be denoted with $\upsilon \in \Theta$.

Theorem 5.2. *The Superposition Operator (SO) is a global ε-accurate approximator of continuously differentiable functions $f : \Theta \rightarrow \Re$.*

Proof. Let us recall the expression of the SO,

$$v = \begin{cases} \sum_{l=0}^{L} w_l S_l(\upsilon; a_l) & \text{if } \upsilon \geq 0 \\ \sum_{l=0}^{-L} w_l S_l(\upsilon; a_l) & \text{if } \upsilon < 0 \end{cases}, \tag{5.29}$$

where $\upsilon \in \Theta$. Let us focus on the case $\upsilon \geq 0$. Each DZO has a slope which is equal to 1. The slope ω_l between υ and v, for $a_l \leq \upsilon \leq a_{l+1}$, is determined by the sum of the weights of each k-th DZO with $a_k < a_{l+1}$, i.e.,

$$\omega_l = \sum_{k=0}^{l} w_l , \tag{5.30}$$

from which one obtains

$$w_l = \omega_l - \omega_{l-1} , \; l = 0 \ldots L . \tag{5.31}$$

Substituting (5.31) into the first equation of (5.29), one obtains

$$v = \sum_{l=0}^{L} (\omega_l - \omega_{l-1}) S_l(\upsilon; a_l) , \; \upsilon \geq 0 . \tag{5.32}$$

Now, let us assume that at t only l DZOs are active (i.e. with a non-zero output). In this case,

$$S_k(\upsilon; a_k) = \upsilon - a_k , \; k = 0 \ldots l , \tag{5.33}$$

and thus (5.32) becomes

$$v = \omega_l \upsilon - \sum_{k=1}^{l} (\omega_k - \omega_{k-1}) a_k . \tag{5.34}$$

Equation (5.34) points out that in each instant the output of the SO is determined by the equation of a line, where ω_l is the slope. To take into account the activation of the DZOs, let us introduce the influence functions $\psi_l(\upsilon)$ which are defined by

$$\psi_l(\upsilon) = \begin{cases} 1 \text{ if } \upsilon \in [a_l, a_{l+1}] \\ 0 \text{ otherwise} \end{cases}, \ l = 0...L, \ a_{L+1} = +\infty . \tag{5.35}$$

Using (5.35), the overall SO can be expressed as

$$v = \sum_{l=0}^{L} \left(\omega_l \upsilon - \sum_{k=1}^{l} (\omega_k - \omega_{k-1}) a_k \right) \psi_l(\upsilon) . \tag{5.36}$$

Since $\omega_l \upsilon - \sum_{k=1}^{l} (\omega_k - \omega_{k-1}) a_k$ is a local ε-accurate approximator [76], and since $\psi_l(\upsilon) \geq 0$ and $\{\psi_l(\upsilon)\}$ satisfy (5.9), theorem 5.1 can be applied to demonstrate that the SO is a global ε-accurate approximator of continuous function defined in the compact set where υ belongs (the co-domain of the PIM). The proof is complete.

5.3. Adaptive control with the MPIM

In Figure 5.3 the adaptive control-loop represented in Figure 5.2 is specialized for the case of a MPIM.

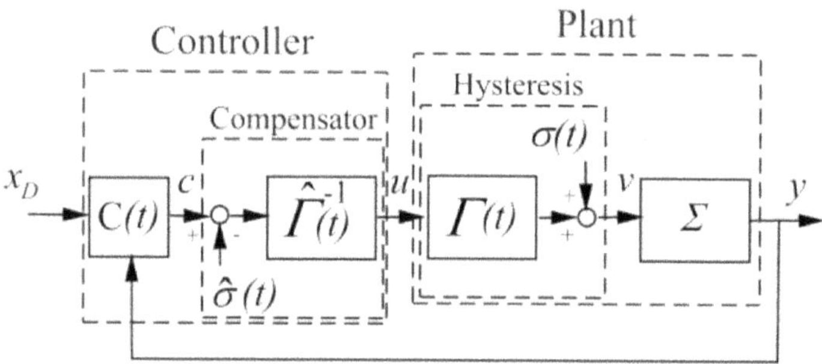

Figure 5.3 – Adaptive control-loop for a MPIM (a time-varying offset $\sigma(t)$ has been added)

The main difference between Figure 5.2 and Figure 5.3 is that in the latter a time-varying offset has been added, since the MPIM cannot entail this effect. Consequently, the overall compensator must subtract the estimated offset $\hat{\sigma}$ from the control variable c. To clarify this detail, let us consider the ideal case where the compensator is the perfect inverse of the hysteresis phenomenon, i.e. $\Gamma[\hat{\Gamma}^{-1}[u]] = u$ and $\hat{\sigma} = \sigma$. The output v of the hysteresis is

$$v = \Gamma[u] + \sigma = \Gamma[\Gamma^{-1}[c - \sigma]] = c - \sigma + \sigma = c, \qquad (5.37)$$

which actually gives the desired result, that is the correspondence between c and v.

The influence of temperature to the control-loop is taken into account through the variation of the parameters of the hysteresis model and the offset. The aim of the adaptive compensator is to trace those variations and offer in each instant the best possible compensation. The controller C acts on the plant via a *feedback linearization strategy*. The strategy is illustrated in the coming section.

5.3.1. The tracking error dynamics

Let us refer to the dynamics of the state x reported in (5.2). The function $f(x)$ is unknown, and will be approximated by a piecewise linear approximator $\hat{f}(x;\hat{\theta})$ expressed in (5.14). The unknown gain g is approximated by the estimated gain $\hat{g}$. It is assumed that the input gain is positive, i.e. $g_0 + g > 0$. The parameter vector $\hat{\theta}$ and the estimated gain $\hat{g}$ are to be adapted on-line, based on the available measurements. In particular, it is assumed that the complete state vector x is measurable and available for calculations. The domain $X \subset \Re^n$ of the unknown function is called the *approximation region.* The desired trajectory is assumed to belong to this region at each time, $x_D \in X$ for all $t \in \Re^+$. The local supports of the basis functions of $\hat{f}(x;\hat{\theta})$ cover X, i.e. $X \subseteq \bigcup_i S_i$.

The control action c is assumed to be

$$c = \frac{1}{g_0 + \hat{g}} \left(\eta_0 - f_0(x) - \hat{f}(x;\hat{\theta}) + x_{D1}^{(n)} - \eta_R \right), \tag{5.38}$$

where η_0 is a contribution which aims at stabilizing the linear part of the plant, $x_{D1}^{(n)}$ is the n-th derivative of the desired displacement trajectory x_{D1}, and η_R is a contribution which enhances the robustness of the control-loop with respect to model uncertainties and disturbances. We notice that η_R can be chosen in several ways. For instance, it can be defined as a bounding control contribution that ensures the asymptotic convergence of the tracking error to zero, or a contribution which acts only in the undesired case that the state x goes outside of the approximation region [77]. The following analysis is carried out in such a way that the particular choice of η_R does not change the stability properties of the closed-loop.

A convenient choice for η_0 is

$$\eta_0 = -K^T \tilde{x}\,, \tag{5.39}$$

where K is the control gains vector, $K = [K_1,\dots,K_n]^T$, and $\tilde{x}$ is the tracking error defined in (5.4). The vector K is chosen such that the matrix $(A - K^T B)$ is a Hurwitz matrix. Let us report the dynamics of the state in (5.2),

$$\dot{x} = A\dot{x} + B\left(f_0(x) + f(x) + (g_0 + g)v\right), \tag{5.40}$$

from which it can be noted that the input to the system is the output v of the unknown hysteresis Γ. Since the estimated compensator $\hat{\Gamma}^{-1}$ is the exact inverse of the corresponding hysteresis model $\hat{\Gamma}$ (a special feature of the MPIM, see section 3.3), it can be written that

$$c = \hat{\Gamma}\left[\hat{\Gamma}^{-1}[c - \hat{\sigma}]\right] + \hat{\sigma} = \hat{\Gamma}[u] + \hat{\sigma}\,, \tag{5.41}$$

and thus

$$v - c = \Gamma[u] - \hat{\Gamma}[u] + \sigma - \hat{\sigma}\,. \tag{5.42}$$

Equation (5.42) points out that the difference between the control action c and v is equal to the difference between the actual hysteresis phenomenon and its estimation. Using (5.23), (5.42) becomes

$$v = c - \left(\tilde{\Gamma} + \tilde{\sigma} - \varepsilon_\Gamma\right), \tag{5.43}$$

where $\tilde{\Gamma} = \tilde{\Gamma}[u] = \hat{\Gamma}[u] - \Gamma^*[u]$, $\tilde{\sigma}$ is as in (5.28), and $\varepsilon_\Gamma = \varepsilon_\Gamma(u;N,L)$. The substitution of (5.43) and (5.38) into (5.40) leads to the following dynamics of the tracking error,

$$\dot{\tilde{x}} = (A - BK^T)\tilde{x} + B\left(f(x) - \hat{f}(x;\hat{\theta}) + (g - \hat{g})c - \eta_R - (g_o + g)\left(\tilde{\Gamma} + \tilde{\sigma} - \varepsilon_\Gamma\right)\right), \tag{5.44}$$

where the term $f(x)-\hat{f}(x;\hat{\theta})$ appears. Since $f(x)-\hat{f}(x;\hat{\theta})=f^*(x)-\hat{f}(x)+\varepsilon_f$, equation (5.44) becomes

$$\dot{\tilde{x}}=(A-BK^T)\tilde{x}+B\left(-\tilde{\theta}^T\phi_f(x)+\varepsilon_f+(g-\hat{g})c-\eta_R-(g_o+g)\left(\tilde{\Gamma}+\tilde{\sigma}-\varepsilon_\Gamma\right)\right), \qquad (5.45)$$

where $\varepsilon_f=\varepsilon_f(x;N_f)$, and $\phi_f(x)$ is defined in (5.14). Equation (5.45) is influenced by the hysteresis error $\tilde{\Gamma}+\tilde{\sigma}-\varepsilon_\Gamma$, which is analyzed in the next section.

5.3.2. The hysteresis error

In this section we focus on the term $\tilde{\Gamma}+\tilde{\sigma}$ which appears in the tracking error dynamics (5.45). It depends on the hysteresis parameter deviations $\tilde{w}$, $\tilde{p}$, $\tilde{r}$, $\tilde{a}$ defined in (5.24)-(5.27). The adaptive compensation strategy adapts the parameters in order to improve the performance of the compensation.

One question is if all of the parameters should be adapted, or only a subset of them. In sections 3.3.2 and 3.3.7 it has been told that usually the thresholds r and a are fixed a priori, and only the weights w and p are identified to fit the hysteresis model to the data. This is certainly a solution, and it means considering the (fixed) parameters $\hat{r}$ and $\hat{a}$ already at their best values, $\hat{r}\equiv r^*$ and $\hat{a}\equiv a^*$. If this assumption holds true with respect to the real plant, then the MFAE ε_Γ will have a certain value; if the assumption does not hold, then the MFAE ε_Γ will be bigger than in the other case.

In this thesis, the choice is to fix r and adapt w, p, a. To reach this goal the MPIM must be expressed in a convenient way [79]. Let us recall that

$$\tilde{\Gamma}=\hat{w}^T\underline{S}^{(M)}\left(\hat{p}^T\underline{H}[u];\hat{a}\right)-w^{*T}\underline{S}^{(G)}\left(p^{*T}\underline{H}[u];a^*\right), \qquad (5.46)$$

where the thresholds r have been omitted since they are not involved in the adaptation, and the initial states are assumed to be zero.

The numbers M and G emphasize the number of elementary DZOs which are active in the estimation $\hat{\Gamma}$ and in the best approximation Γ^*, respectively. The vectors $\underline{S}^{(M)}(\hat{p}^T \underline{H}[u];\hat{a})$ and $\underline{S}^{(G)}(p^{*T} \underline{H}[u];a^*)$ have a different number of DZOs that have a nonzero output, because $\hat{p} \neq p^*$ and $\hat{a} \neq a^*$. Since the number of DZOs that are active in the best approximation is not known, the difference in (5.46) can be evaluated only on those operators that are known to be active, i.e. the DZOs which are active in the estimation $\underline{S}^{(M)}(\hat{p}^T \underline{H}[u];\hat{a})$. Equation (5.46) can be rewritten as

$$\tilde{\Gamma} = \hat{w}^T \underline{S}^{(M)}\left(\hat{p}^T \underline{H}[u];\hat{a}\right) - w^{*T} \underline{S}^{(M)}\left(p^{*T} \underline{H}[u];a^*\right) + \varepsilon_A, \tag{5.47}$$

where the difference is now evaluated on the same M active operators. Due to this choice, the activation error $\varepsilon_A = \varepsilon_A(\tilde{p},\tilde{a})$ appears, which depends on the deviation of $\hat{p}$ and $\hat{a}$ from their best values. If the deviations are small, the activation error will be small too. Good starting values of the parameters, offered by an off-line identification, can in fact realize this situation [80], [79]. Hereafter the superscript M is omitted for ease of notation.

The dependence of the MPIM on the thresholds of the SO can be expressed in an alternative way. Let us define the vector of distances Δ,

$$\Delta = [\Delta_{-L},\ldots,\Delta_{-1},\Delta_0 = 0,\Delta_1,\ldots,\Delta_L]^T \in \Re^{2L+1}, \tag{5.48}$$

such that the following relationships hold

$$a_l = \sum_{k=0}^{l} \Delta_k, \quad a_l \geq 0, \tag{5.49}$$

$$a_l = \sum_{k=0}^{-l} \Delta_k \,, \; a_l \le 0 \,, \tag{5.50}$$

with $\Delta_0 = 0$. It is easy to show that

$$a = W\Delta \,, \tag{5.51}$$

where the matrix W is defined in (3.52). In order for (5.49) and (5.50) to hold, Δ must satisfy the constraints

$$\begin{bmatrix} -\underline{I}^{L\times L} & \underline{0}^{L\times(L+1)} \\ \underline{0}^{(L+1)\times(L)} & \underline{I}^{(L+1)\times(L+1)} \end{bmatrix} \Delta \ge 0 \,, \tag{5.52}$$

where $\underline{I}^{n\times m}$ and $\underline{0}^{n\times m}$ are, respectively, the identity matrix and the zero matrix with the given dimensions. The inequality (5.52) simply states that $\Delta_l \le 0$ for $l = -L \ldots -1$, and that $\Delta_l \ge 0$ for $l = 0 \ldots L$. Recall that $\Delta_0 = 0$.

Thanks to (5.48)-(5.50), the difference (5.47) can be written as

$$\tilde{\Gamma} = \hat{w}^T \underline{S}\left(\hat{p}^T \underline{H}[u]; \hat{\Delta}\right) - w^{*T} \underline{S}\left(p^{*T} \underline{H}[u]; \Delta^*\right) + \varepsilon_A \,, \tag{5.53}$$

where $\hat{\Delta}$ is the estimation of the unknown best value Δ^*. The Taylor series expansion of (5.53) around the point $(\hat{w}, \hat{p}, \hat{\Delta})$ leads to

$$\tilde{\Gamma} = \tilde{w}^T \underline{S}\left(\hat{p}^T \underline{H}[u]; \hat{\Delta}\right) + \hat{w}^T \frac{\partial \underline{S}\left(\hat{p}^T \underline{H}[u]; \hat{\Delta}\right)}{\partial p} \tilde{p} + \hat{w}^T \frac{\partial \underline{S}\left(\hat{p}^T \underline{H}[u]; \hat{\Delta}\right)}{\partial \Delta} \tilde{\Delta} + \varepsilon_T + \varepsilon_A \,, \tag{5.54}$$

where $\tilde{\Delta} = \hat{\Delta} - \Delta^*$ and ε_T is an error due to the truncation of the series (see section 5.2.4 for the meaning of the other symbols).

The introduction of Δ makes the development of a *projection algorithm* much easier. This is an important issue also for w and p. They must satisfy (3.50) in order for the MPIM to be invertible. The projection algorithm has to ensure the invertibility during adaptation. The development of such an algorithm becomes easier if the alternative parameters

$$\omega = W^{-1} w , \tag{5.55}$$

$$\rho = P^{-1} p , \tag{5.56}$$

are considered instead of w and p. With (5.55) and (5.56), (5.54) becomes

$$\tilde{\Gamma} = \tilde{\omega}^T W^{-T} \underline{S}\left(\hat{p}^T \underline{H}[u];\hat{\Delta}\right) + \tilde{\rho}^T P^{-T} S_p^T \hat{w} + \tilde{\Delta}^T S_\Delta^T \hat{w} + \varepsilon_T + \varepsilon_A , \tag{5.57}$$

where $S_p = \partial \underline{S}\left(\hat{p}^T \underline{H}[u];\hat{\Delta}\right) / \partial p$, $S_\Delta = \partial \underline{S}\left(\hat{p}^T \underline{H}[u];\hat{\Delta}\right) / \partial \Delta$, $\tilde{\omega} = \hat{\omega} - \omega^*$, $\tilde{\rho} = \hat{\rho} - \rho^*$.

We are now ready to express the hysteresis error as

$$\tilde{\Gamma} + \tilde{\sigma} = \tilde{\omega}^T W^{-T} \underline{S}\left(\hat{p}^T \underline{H}[u];\hat{\Delta}\right) + \tilde{\rho}^T P^{-T} S_p^T \hat{w} + \tilde{\Delta}^T S_\Delta^T \hat{w} + \varepsilon_T + \varepsilon_A + \tilde{\sigma} , \tag{5.58}$$

which is suitable for the development of an adaptive algorithm.

5.3.3. Stability analysis of the control-loop

Equation (5.58) leads to the following tracking error dynamics

$$\begin{aligned} \dot{\tilde{x}} = {} & (A - BK^T)\tilde{x} + B\left(-\tilde{\theta}^T \phi_f(x) + \varepsilon_f - \tilde{g}c - \eta_R\right) + \\ & - B(g_o + g)\left(\tilde{\omega}^T W^{-T} \underline{S}\left(\hat{p}^T \underline{H}[u];\hat{\Delta}\right) + \tilde{\rho}^T P^{-T} S_p^T \hat{w} + \tilde{\Delta}^T S_\Delta^T \hat{w} + \varepsilon_1 + \tilde{\sigma}\right) \end{aligned} , \tag{5.59}$$

where $\varepsilon_1 = \varepsilon_T + \varepsilon_A - \varepsilon_\Gamma$ and $\tilde{g} = \hat{g} - g$.

To determine the behavior of the tracking error, let us consider the candidate Lyapunov function

$$V = \tilde{x}^T L \tilde{x} + \frac{1}{\gamma_\theta}\tilde{\theta}^T\tilde{\theta} + \frac{1}{\gamma_\rho}\tilde{\rho}^T\tilde{\rho} + \frac{1}{\gamma_\omega}\tilde{\omega}^T\tilde{\omega} + \frac{1}{\gamma_\Delta}\tilde{\Delta}^T\tilde{\Delta} + \frac{1}{\gamma_\sigma}\tilde{\sigma}^2 + \frac{1}{\gamma_g}\tilde{g}^2, \quad (5.60)$$

where L is a symmetric positive definite matrix, and γ_θ, γ_ρ, γ_ω, γ_Δ, γ_σ are positive coefficients. The derivative of (5.60) along the trajectories of the system is

$$\begin{aligned}\dot{V} &= 2\tilde{x}^T L\Big((A - BK^T)\tilde{x} + B\big(-\tilde{\theta}^T\phi_f(x) + \varepsilon_f - \tilde{g}c - \eta_R\big) \\ &\quad -B(g_o + g)\Big(\tilde{\omega}^T W^{-T}\underline{S}\big(\hat{\rho}^T\underline{H}[u];\hat{\Delta}\big) + \tilde{\rho}^T P^{-T} S_p^T\hat{w} + \tilde{\Delta}^T S_\Delta^T\hat{w} + \varepsilon_1 + \tilde{\sigma}\Big)\Big). \\ &\quad + \frac{2}{\gamma_\theta}\tilde{\theta}^T\dot{\hat{\theta}} + \frac{2}{\gamma_\rho}\tilde{\rho}^T\dot{\hat{\rho}} + \frac{2}{\gamma_\omega}\tilde{\omega}^T\dot{\hat{\omega}} + \frac{2}{\gamma_\Delta}\tilde{\Delta}^T\dot{\hat{\Delta}} + \frac{2}{\gamma_\sigma}\tilde{\sigma}\dot{\hat{\sigma}} + \frac{2}{\gamma_g}\tilde{g}\dot{\hat{g}}\end{aligned} \quad (5.61)$$

The fact that the matrix $(A - BK^T)$ is a Hurwitz matrix implies that, for each positive definite matrix Q, there exists a positive definite matrix L such that

$$L(A - BK^T) + (A - BK^T)^T L = -Q. \quad (5.62)$$

Let us assume, without loss of generality, that $Q = I$ (the identity matrix).

Equation (5.62) reduces (5.61) to

$$\begin{aligned}\dot{V} &= -\tilde{x}^T\tilde{x} - 2\tilde{\theta}^T\phi_f(x)e_T - 2\tilde{g}ce_T - 2\eta_R e_T + 2\varepsilon_f e_T \\ &\quad -2(g_o + g)\Big(\tilde{\omega}^T W^{-T}\underline{S}\big(\hat{\rho}^T\underline{H}[u];\hat{\Delta}\big) + \tilde{\rho}^T P^{-T} S_p^T\hat{w} + \tilde{\Delta}^T S_\Delta^T\hat{w} + \varepsilon_1 + \tilde{\sigma}\Big)e_T, \\ &\quad + \frac{2}{\gamma_\theta}\tilde{\theta}^T\dot{\hat{\theta}} + \frac{2}{\gamma_\rho}\tilde{\rho}^T\dot{\hat{\rho}} + \frac{2}{\gamma_\omega}\tilde{\omega}^T\dot{\hat{\omega}} + \frac{2}{\gamma_\Delta}\tilde{\Delta}^T\dot{\hat{\Delta}} + \frac{2}{\gamma_\sigma}\tilde{\sigma}\dot{\hat{\sigma}} + \frac{2}{\gamma_g}\tilde{g}\dot{\hat{g}}\end{aligned} \quad (5.63)$$

and thus

$$\dot{V} = -\tilde{x}^T\tilde{x} + \frac{2\tilde{\theta}^T}{\gamma_\theta}\left(\dot{\hat{\theta}} - \gamma_\theta\phi_f(x)e_T\right) + \frac{2\tilde{\omega}^T}{\gamma_\omega}\left(\dot{\hat{\omega}} - \gamma_\omega(g_o+g)W^{-T}\underline{S}\left(\hat{p}^T\underline{H}[u];\hat{\Delta}\right)e_T\right)$$
$$+\frac{2\tilde{\rho}^T}{\gamma_\rho}\left(\dot{\hat{\rho}} - \gamma_\rho(g_o+g)P^{-T}S_p^T\hat{w}e_T\right) + \frac{2\tilde{\Delta}^T}{\gamma_\Delta}\left(\dot{\hat{\Delta}} - \gamma_\Delta(g_o+g)S_\Delta^T\hat{w}e_T\right) \qquad , \qquad (5.64)$$
$$+\frac{2\tilde{\sigma}}{\gamma_\sigma}\left(\dot{\hat{\sigma}} - \gamma_\sigma(g_0+g)e_T\right) + \frac{2\tilde{g}}{\gamma_g}\left(\dot{\hat{g}} - \gamma_g c e_T\right) + 2(\varepsilon - \eta_R)e_T$$

where $e_T = B^T L\tilde{x}$ is called the (scalar) *training error*, since it is used to train the adaptive hysteresis and the adaptive approximator, and $\varepsilon = (-(g_0+g)\varepsilon_1 + \varepsilon_f)$ is the overall approximation error.

Equation (5.64) is rather standard in the adaptive control literature [77]. In the absence of uncertainties and approximation errors ($\varepsilon = 0$), the adaptive laws of the parameters can be obtained by simply zeroing all the brackets of (5.64). In this ideal case, the derivative of the Lyapunov function is $\dot{V} = -\tilde{x}^T\tilde{x} \le 0$, and thus the stability of the control-loop and the asymptotic convergence of the tracking error to zero is ensured. Unfortunately, there are three important issues which complicate the stability analysis: 1) the convergence is negatively influenced by the presence of uncertainties, modeling errors and measurement noise (i.e. ε); 2) the overall approximation error grows when the state is outside of the approximation region X; 3) all of the adapted parameters must remain in proper feasible sets during adaptation.

The presence of modeling errors within X can be taken into account by defining a dead-zone function which turns the adaptation of the parameters off when the tracking error is smaller than a certain value. Let us denote the upper bound to the overall error ε with ε_0, such that

$$\varepsilon_0 \ge \sup_{x\in X}|\varepsilon| \qquad (5.65)$$

holds for each $x \in X$. Let us introduce the dead-zone function

$$\mathrm{dz}\left(e_T, \tilde{x}; \bar{\varepsilon}\right) = \begin{cases} 0 & \text{if } \tilde{x}^T L \tilde{x} \le \bar{\lambda}_L \bar{\varepsilon}^2 \\ e_T & \text{if } \tilde{x}^T L \tilde{x} > \bar{\lambda}_L \bar{\varepsilon}^2 \end{cases}, \tag{5.66}$$

where the dead-zone parameter $\bar{\varepsilon}$ is chosen as

$$\bar{\varepsilon} = 2\|LB\|_2 \varepsilon_0 + \mu , \tag{5.67}$$

where $\mu > 0$ is an arbitrarily small design constant. The dead-zone function (5.66) is extremely useful to prevent parameter drifts since it stops the adaptation when the tracking error goes within a bound determined by $\bar{\varepsilon}$ and the matrix L. In fact, $\bar{\lambda}_L$ indicates the maximum eigenvalue of L. Note that, since L is positive definite, the maximum and the minimum eigenvalues, $\bar{\lambda}_L$ and $\underline{\lambda}_L$, are real positive numbers. The way that the dead-zone function influences the stability analysis is shown later.

Outside of the approximation region, the MFAE of f can become bigger since the approximation does not work, $\hat{f}(x;\theta) = 0$ (because $\phi_f(x) = 0$ for $x \notin X$). The situation can be described by considering

$$f(x) - \hat{f}(x;\theta^*) = \varepsilon_f + E_f \text{, for } x \in \Re^n - X , \tag{5.68}$$

where an additive unknown term E_f appears when the state is outside X. Consequently, the overall approximation error becomes $\varepsilon + E_f$ for $x \in \Re^n - X$. Let us assume that some bounds on E_f are known,

$$b_L \le E_f \le b_U \text{, for } x \in \Re^n - X , \tag{5.69}$$

and let the control term η_R be defined as

$$\eta_R = \begin{cases} \begin{cases} b_U & \text{if } e_T \geq 0 \\ b_L & \text{if } e_T < 0 \end{cases} & \text{if } x \in \Re^n - X \\ 0 & \text{if } x \in X \end{cases} . \tag{5.70}$$

This control term counteracts the effects of the increased approximation errors outside X. Its efficacy is shown in the stability analysis. Note that $(E_f - \eta_R)e_T \leq 0$.

The use of projection algorithms can indeed solve the third issue, ensuring that the adapted parameters remain in their feasible sets. Let us recall (section 3.3.7) that the parameters $\hat{\omega}$, $\hat{\rho}$ of the hysteresis model must satisfy

$$\hat{\omega} \geq \delta_W , \tag{5.71}$$

$$\hat{\rho} \geq \delta_P , \tag{5.72}$$

which come directly from (3.50). The inequalities (5.71) and (5.72) emphasize that each weight is only lower bounded by δ. Something similar occurs for $\hat{\Delta}$, which must satisfy (5.52). We introduce the following projection algorithm for $\hat{\rho}$ and $\hat{\omega}$,

$$\text{prj}(\dot{\chi}) = \begin{cases} \dot{\chi} \ (\text{if } \chi > \delta) \ \text{ or } \ (\chi = \delta \text{ and } \dot{\chi} \geq 0) \\ 0 \ \text{otherwise} \end{cases} , \tag{5.73}$$

where χ refers to the generic component of the vectors $\hat{\rho}$ and $\hat{\omega}$, and the following projection algorithm for Δ,

$$\mathrm{prj}_{\Delta}(\dot{\varsigma}) = \begin{cases} \begin{cases} \dot{\varsigma} \ (\text{if } \varsigma < 0) \ \text{or} \ (\varsigma = 0 \ \text{and} \ \dot{\varsigma} \le 0) \\ 0 \ \text{otherwise} \end{cases} & \text{if } \varsigma \in [\Delta_{-L}, \Delta_{-1}] \\ 0 \ \text{if } \varsigma = \Delta_0 & \\ \begin{cases} \dot{\varsigma} \ (\text{if } \varsigma > 0) \ \text{or} \ (\varsigma = 0 \ \text{and} \ \dot{\varsigma} \ge 0) \\ 0 \ \text{otherwise} \end{cases} & \text{if } \varsigma \in [\Delta_1, \Delta_L] \end{cases}, \tag{5.74}$$

where ς refers to the generic component of the vector Δ. The projection (5.74) ensures that the thresholds of the negative DZOs remain negative, and that the thresholds of the positive DZOs remain positive. The threshold $a_0 = \Delta_0 = 0$ is not changed during adaptation to preserve the presence of the linear operator.

Another parameter which must be constrained to the admissible set is $\hat{g}$, that is the estimation of the unknown input gain of the system Σ. It has to be noted that $1/(g_0 + \hat{g})$ appears in the expression of the control action (5.38), and thus

$$\hat{g} \ge -g_0 + \mu \tag{5.75}$$

must hold at any time, with $\mu > 0$ being a small design constant. The following projection algorithm is introduced for $\hat{g}$:

$$\mathrm{prj}_g\left(\dot{\hat{g}}\right) = \begin{cases} \dot{\hat{g}} \ (\text{if } \hat{g} > -g_0 + \mu) \ \text{or} \ \left(\hat{g} = -g_0 + \mu \ \text{and} \ \dot{\hat{g}} \ge 0\right) \\ 0 \ \text{otherwise} \end{cases}. \tag{5.76}$$

If the dead-zone function (5.66) is used, and if the projection algorithms (5.73)-(5.76) are applied to the adaptation laws of the corresponding parameters, it is ensured that the adaptation stops when the error is within a certain bound and that the adapted parameters remain in their feasible sets. It is necessary to analyze what is the influence of those algorithms on the stability of the closed-loop system, which basically depends on the Lyapunov derivative (5.64).

Based on (5.64), the following adaptive laws for the parameters can be introduced,

$$\dot{\hat{\theta}} = \gamma_\theta \phi_f(x)\mathrm{dz}\left(e_T, \tilde{x}; \overline{\varepsilon}\right), \tag{5.77}$$

$$\dot{\hat{\omega}} = \mathrm{prj}\left(\gamma_\omega (g_o + g) W^{-T} \underline{S}\left(\hat{p}^T \underline{H}[u]; \hat{\Delta}\right)\mathrm{dz}\left(e_T, \tilde{x}; \overline{\varepsilon}\right)\right), \tag{5.78}$$

$$\dot{\hat{p}} = \mathrm{prj}\left(\gamma_\rho (g_o + g) P^{-T} S_p^T \hat{w}\mathrm{dz}\left(e_T, \tilde{x}; \overline{\varepsilon}\right)\right), \tag{5.79}$$

$$\dot{\hat{\Delta}} = \mathrm{prj}_\Delta\left(\gamma_\Delta (g_o + g) S_\Delta^T \hat{w}\mathrm{dz}\left(e_T, \tilde{x}; \overline{\varepsilon}\right)\right), \tag{5.80}$$

$$\dot{\hat{\sigma}} = \gamma_\sigma (g_0 + g)\mathrm{dz}\left(e_T, \tilde{x}; \overline{\varepsilon}\right), \tag{5.81}$$

$$\dot{\hat{g}} = \mathrm{prj}_g\left(\gamma_g c\mathrm{dz}\left(e_T, \tilde{x}; \overline{\varepsilon}\right)\right), \tag{5.82}$$

where the unknown term $(g_o + g)$ appears. However, since γ_ρ, γ_ω, γ_Δ and γ_σ are arbitrary positive, then the corresponding products $\gamma_\omega(g_o + g)$, $\gamma_\rho(g_o + g)$, $\gamma_\Delta(g_o + g)$, $\gamma_\sigma(g_0 + g)$ can also be determined in the design phase, and are the *adaptive gains* of the parameters.

The theorem that follows determines the stability properties of the control-loop as well as the performance that can be achieved on the tracking error.

Theorem 5.3. *Let us refer to the adaptive closed-loop (5.2),with the control action c specified in (5.38). Let the robustness term η_R be defined as in* (5.70). *If the parameters of the adaptive hysteresis model and the adaptive controller are adapted with (5.77)-(5.82), then the tracking error $\tilde{x}$, the training error e_T and the deviations $\tilde{\theta}$, $\tilde{\omega}$, $\tilde{p}$, $\tilde{g}$, $\tilde{\sigma}$, $\tilde{\Delta}$ are bounded and belong to their respective feasible sets; moreover, the Euclidean norm of the tracking error is ultimately bounded by $\overline{\varepsilon}$, i.e. $\|\tilde{x}\|_2 \to [0, \overline{\varepsilon}]$ when $t \to +\infty$.*

Proof. We first show what happens to the Lyapunov derivative in (5.64) when the dead-zone function is not active ($\mathrm{dz}\left(e_T,\tilde{x};\overline{\varepsilon}\right)=e_T$), independently of wheter $x\in X$ or $x\in\Re^n-X$. Let us define the time intervals (t_{s_i},t_{f_i}) so that the condition $\tilde{x}^T L\tilde{x}>\overline{\lambda}_L\overline{\varepsilon}^2$ is satisfied only for $t\in(t_{s_i},t_{f_i})$, with $i=1,2,3...$, and $t_{s_i}<t_{f_i}\le t_{s_{i+1}}$. Briefly speaking, when $t\in(t_{s_i},t_{f_i})$ then $\mathrm{dz}\left(e_T,\tilde{x};\overline{\varepsilon}\right)=e_T$; when $t\in[t_{f_i},t_{s_{i+1}}]$ then $\mathrm{dz}\left(e_T,\tilde{x};\overline{\varepsilon}\right)=0$. Since

$$\tilde{x}(t_{f_i})^T L\tilde{x}(t_{f_i})=\tilde{x}(t_{s_{i+1}})^T L\tilde{x}(t_{s_{i+1}})=\overline{\lambda}_L\overline{\varepsilon}^2 \tag{5.83}$$

and parameter estimations (5.77)-(5.82) are turned off when $t\in[t_{f_i},t_{s_{i+1}}]$, then for the Lyapunov function in (5.60) one obtains

$$V(t_{f_i})=V(t_{s_{i+1}})\,. \tag{5.84}$$

When $t\in(t_{s_i},t_{f_i})$, then $\tilde{x}^T L\tilde{x}>\overline{\lambda}_L\overline{\varepsilon}^2$, and by use of (5.67) one obtains

$$\tilde{x}^T L\tilde{x}>\overline{\lambda}_L\left(2\|LB\|_2\,\varepsilon_0+\mu\right)^2,\ \text{for}\ t\in(t_{s_i},t_{f_i})\,. \tag{5.85}$$

Exploiting the property $\tilde{x}^T L\tilde{x}\le\overline{\lambda}_L\|\tilde{x}\|_2^2$, it comes that

$$\|\tilde{x}\|_2>2\|LB\|_2\,\varepsilon_0+\mu\ ,\ \text{for}\ t\in(t_{s_i},t_{f_i})\,. \tag{5.86}$$

Thus, when $t\in(t_{s_i},t_{f_i})$ and when $x\in X$, the Lyapunov derivative (5.64) becomes

$$\dot{V} = -\tilde{x}^T\tilde{x} + \frac{2\tilde{\theta}^T}{\gamma_\theta}\left(\dot{\hat{\theta}} - \gamma_\theta\phi_f(x)e_T\right) + \frac{2\tilde{\omega}^T}{\gamma_\omega}\left(\dot{\hat{\omega}} - \gamma_\omega(g_o+g)W^{-T}\underline{S}\left(\hat{p}^T\underline{H}[u];\hat{\Delta}\right)e_T\right)$$
$$+\frac{2\tilde{\rho}^T}{\gamma_\rho}\left(\dot{\hat{\rho}} - \gamma_\rho(g_o+g)P^{-T}S_p^T\hat{w}e_T\right) + \frac{2\tilde{\Delta}^T}{\gamma_\Delta}\left(\dot{\hat{\Delta}} - \gamma_\Delta(g_o+g)S_\Delta^T\hat{w}e_T\right) \quad , \tag{5.87}$$
$$+\frac{2\tilde{\sigma}}{\gamma_\sigma}\left(\dot{\hat{\sigma}} - \gamma_\sigma(g_0+g)e_T\right) + \frac{2\tilde{g}}{\gamma_g}\left(\dot{\hat{g}} - \gamma_g c e_T\right) + 2\tilde{x}^T LB\varepsilon$$

where in the last term the definition of the training error, $e_T = \tilde{x}^T LB$, has been used. Substituting the adaptive laws (5.77)-(5.82) into (5.87) and assuming that projection is not active, one obtains

$$\begin{aligned}\dot{V} &= -\tilde{x}^T\tilde{x} + 2\tilde{x}^T LB\varepsilon \\ &\le -\|\tilde{x}\|_2^2 + 2\|\tilde{x}\|_2\|LB\|_2\varepsilon_0 \\ &\le -\|\tilde{x}\|_2\left(\|\tilde{x}\|_2 - 2\|LB\|_2\varepsilon_0\right) \\ &\le -\bar{\varepsilon}\mu\end{aligned} \text{, for } t\in(t_{s_i},t_{f_i}) \,. \tag{5.88}$$

When the state is outside X , the Lyapunov derivative (5.64) is instead

$$\dot{V} = -\tilde{x}^T\tilde{x} + 2\tilde{x}^T LB\varepsilon + 2\left(E_f - \eta_R\right)e_T , \tag{5.89}$$

where the term $(E_f - \eta_R)e_T$ is always negative semidefinite thanks to (5.70). Thus, also in this case (5.89) becomes

$$\dot{V} \le -\tilde{x}^T\tilde{x} + 2\tilde{x}^T LB\varepsilon \le -\bar{\varepsilon}\mu \text{, for } t\in(t_{s_i},t_{f_i}) \,. \tag{5.90}$$

Equations (5.88) and (5.90) can be summarized as

$$\dot{V} \le -\bar{\varepsilon}\mu \text{, for } x\in\Re^n \text{ and } t\in(t_{s_i},t_{f_i}) \,. \tag{5.91}$$

Consequently, $V(t_{f_i}) \le V(t_{s_i})$ is always verified. Integrating both sides of (5.91) for $t \in (t_{s_i}, t_{f_i})$, one obtains

$$V(t_{f_i}) \le V(t_{s_i}) - \bar{\varepsilon}\mu(t_{f_i} - t_{s_i}) . \tag{5.92}$$

However, since (5.84) holds for all $i \in \mathbb{N}^+ - \{0\}$, (5.92) can be completed as follows

$$\begin{aligned} V(t_{f_i}) &\le V(t_{s_i}) - \bar{\varepsilon}\mu(t_{f_i} - t_{s_i}) \le V(t_{f_{i-1}}) - \bar{\varepsilon}\mu(t_{f_i} - t_{s_i}) \\ &\le V(t_{f_{i-2}}) - \bar{\varepsilon}\mu\left((t_{f_i} - t_{s_i}) + (t_{f_{i-1}} - t_{s_{i-1}})\right) \\ &\le \ldots \\ &\le V(0) - \bar{\varepsilon}\mu\sum_{k=1}^{i}\left(t_{f_k} - t_{s_k}\right) \end{aligned} , \tag{5.93}$$

which leads to

$$\sum_{k=1}^{i}\left(t_{f_k} - t_{s_k}\right) \le \frac{V(0)}{\bar{\varepsilon}\mu} . \tag{5.94}$$

Note that the intervals $t \in (t_{s_i}, t_{f_i})$ represent the period of times that $\tilde{x}^T L \tilde{x} > \bar{\lambda}_L \bar{\varepsilon}^2$. The inequality (5.94) shows that the total amount of time that $\tilde{x}^T L \tilde{x} > \bar{\lambda}_L \bar{\varepsilon}^2$ is then finite. This implies that the tracking error evolves in such a way that $\|\tilde{x}\|_2 \to [0, \bar{\varepsilon}]$. Since the parameters are adapted only when $\tilde{x}^T L \tilde{x} > \bar{\lambda}_L \bar{\varepsilon}^2$, inequality (5.94) also tells that the parameter adaptation will stop after a certain amount of time; moreover, thanks to the projection algorithms, the parameters $\tilde{\theta}$, $\tilde{\omega}$, $\tilde{\rho}$, $\tilde{g}$, $\tilde{\sigma}$, $\tilde{\Delta}$ must be bounded.

So far, it has been assumed that the projection algorithms were not active. The last part of the proof concerns the case where the projection algorithm of some parameters is active. The presence of the projection ensures that the parameters remain in their feasible sets. Let us assume, without loss of generality, that only the adaptation of ω_k is zero because of projection. The following conditions are verified: $\hat{\omega}_k = \delta$, $\dot{\hat{\omega}}_k = \text{proj}(\dot{\chi}) = 0$ and $\dot{\chi} < 0$ (see the projection algorithm in (5.73)). The contribution to the Lyapunov derivative (5.64) due to ω_k is

$$\frac{2\tilde{\omega}_k}{\gamma_\omega}\left(0 - \left(\gamma_\omega(g_o + g)W^{-T}\underline{S}\left(\hat{p}^T\underline{H}[u];\hat{\Delta}\right)\mathrm{dz}\left(e_T, \tilde{x}; \overline{\varepsilon}\right)\right)_k\right), \tag{5.95}$$

where the notation $(\cdot)_k$ is used to refer to the k-th row of the matrix within the brackets. From the adaptation law (5.78) we have

$$\dot{\chi} = \left(\gamma_\omega(g_o + g)W^{-T}\underline{S}\left(\hat{p}^T\underline{H}[u];\hat{\Delta}\right)\mathrm{dz}\left(e_T, \tilde{x}; \overline{\varepsilon}\right)\right)_k < 0, \tag{5.96}$$

and follows that $-(\gamma_\omega(g_o + g)W^{-T}\underline{S}(\hat{p}^T\underline{H}[u];\hat{\Delta})\mathrm{dz}(e_T, \tilde{x}; \overline{\varepsilon}))_k > 0$. It is assumed that the best value ω_k^* belongs to the admissible set, i.e. $\omega_k^* \geq \delta$. As a consequence, the first factor of (5.95) is $\tilde{\omega}_k = \hat{\omega}_k - \omega_k^* = \delta - \omega_k^* \leq 0$. Thus, the contribution (5.95) given by the projection (5.73) to the Lyapunov derivative (5.64), is negative semidefinite. This implies that the conclusion (5.91) does not change when the projection is active. A similar discussion can be made for any projection algorithm on any parameter. Hence, the proof of theorem 5.3 is complete.

5.3.4. Summary of the control algorithm

At this point, the control strategy has been discussed in detail together with the adaptation laws of the hysteresis model. Here, we summarize the steps of the control strategy to clarify the previous discussion. At time t, the control algorithm does:

I. Calculation of the tracking error $\tilde{x}$ and of the training error e_T.

II. Calculation of the dead-zone function (5.66).

III. Calculation of the robustness term η_R in (5.70).

IV. Calculation of the adaptation terms (5.77)-(5.82) with projection. This step gives the new estimations $\hat{\theta}$, $\hat{\omega}$, $\hat{\rho}$, $\hat{g}$, $\hat{\sigma}$, $\hat{\Delta}$.

V. The value of $\hat{\theta}$ is used to calculate the approximation $\hat{f}(x;\hat{\theta})$.

VI. The values of $\hat{\rho}$, $\hat{\omega}$ and $\hat{\Delta}$ are used to calculate the parameters of the MPIM p, w and a through (5.56), (5.55) and (5.51).

VII. The MPIM is used to calculate the corresponding compensator IMPIM, using (3.32), (3.33) and (3.39)-(3.43).

VIII. The control action c is calculated as in (5.38).

IX. The estimated offset $\hat{\sigma}$ is subtracted from c.

X. The signal $c-\hat{\sigma}$ enters the estimated IMPIM. The output u, calculated as in (3.45), goes to the plant (refer to Figure 5.3).

The adaptive algorithm described in I-IX can be improved or changed in several ways, for example by adding more terms to the control action c in order to improve the tracking performance, or to take into account other effects such as input saturation. A detailed discussion of these possibilities is beyond the scope of this thesis, since they rely on results which are available in the adaptive control literature. The main contribution of the present section is to show how it is possible to "plug" an adaptive hysteresis component into the existing analytical frameworks of adaptive control. For this reason, the next section presents a simple case where the dynamics of the plant are linear and of the first order. This allows focusing on the advantages of the adaptive hysteresis, and especially the advantages of adapting also the thresholds of the SO.

5.4. Adaptive control of first order linear systems

Referring to the control-loop of Figure 5.3, we consider here the case that Σ is a first order filter with a unity DC gain,

$$\Sigma : \begin{cases} \dot{x} = -gx + gv \\ y = x \\ x(0) \end{cases}, \tag{5.97}$$

where $v \in \Re$ is the output of an unknown hysteresis operator and $g > 0$. We assume $g_0 = 0$ and $f_0(x) \equiv 0$. The tracking error is now $\tilde{x} = x - x_D$. It is assumed that the derivative $x_D^{(1)}$ is available and continuous. The training error coincides with the tracking error, $e_T = \tilde{x}$. The approximation region is $X = \Re$, and thus $\eta_R = 0$. An alternative choice of η_R for controlling MSM actuators can be found in [80], [64]. The control action c becomes now

$$c = \frac{1}{\hat{g}}\left(-K_P\tilde{x} + \hat{g}x + x_D^{(1)}\right), \tag{5.98}$$

where $\eta_0 = -K_P\tilde{x}$ is a proportional action and $-\hat{f}(x;\hat{\theta}) = \hat{g}x$ in this case. The dead zone function can be re-defined as

$$\mathrm{dz}\left(\tilde{x};\overline{\varepsilon}\right) = \begin{cases} 0 \text{ if } \tilde{x} \le \overline{\varepsilon} \\ \tilde{x} \text{ if } \tilde{x} > \overline{\varepsilon} \end{cases}, \text{ with } \overline{\varepsilon} = \frac{\varepsilon_0}{K_P} + \mu \tag{5.99}$$

where $\varepsilon_0 = \sup|\varepsilon|$ and ε represents the overall approximation errors as before. The adaptive laws do not change except for the adaptation of $\hat{g}$, which becomes

$$\dot{\hat{g}} = \mathrm{prj}_g\left(\gamma_g(x - c)\mathrm{dz}\left(\tilde{x};\overline{\varepsilon}\right)\right), \tag{5.100}$$

with the same projection in (5.76).

By means of the same mathematical steps of section 5.3.3, it is possible to prove that the tracking error is ultimately bounded by $\bar{\varepsilon}$. It can be noted that increasing K_P (the proportional action) decreases the ultimate bound of the tracking error. However, this is not a desired operation, since increasing the proportional gain could excite unmodeled dynamics of the plant. Usually, the designer should try to reduce the approximation error by enhancing the approximation capabilities of the adaptive approximators.

5.4.1. Simulation results

Due to several factors such as measurement noise, in the experimental results that follow in section 5.5 the adaptation of the thresholds has been set to zero by choosing $\gamma_\Delta = 0$ since it was not improving the tracking result. Here, some simulation results show that the adaptation of the thresholds is, in general, an advantage.

The plant is now simulated by the series of the MPIM hysteresis Γ and the linear first-order dynamics with $g = 120$. The hysteresis Γ has been obtained as follows. First, the hysteresis of the actuator in Figure 2.4 has been identified based on the experimental data (the obtained curve is already shown in Figure 4.7). Let us call with w_0, p_0, a_0, r_0 the identified values. After, the identified values have been modified into w, p, a, r_0, and those new values have been used to simulate the plant hysteresis Γ. The adaptive compensator, instead, has been initialized with w_0, p_0, a_0, r_0, in order to simulate the non-optimal knowledge of the hysteresis. With the expression "real values" we refer to w, p, a, r_0 used to simulate the "real" plant. The tracking results of two controllers are compared: one uses an adaptive compensator which does not adapt the thresholds (i.e. $\gamma_\Delta = 0$), whereas the other one adapts also the thresholds.

We consider two scenarios. In the first, the alteration of the thresholds of the MPIM is small: in particular, the weights and thresholds have been taken as $[w, p] = 0.8[w_0, p_0]$, $r = r_0$ and $a = 0.8a_0$. The values of the gain K_P and of the other adaptive gains are exactly the same for both schemes. Figure 5.4 shows a zoom of the tracking of a sum of two sinusoids, one at 1 Hz and the other one at 10 Hz. The adaptive controller with thresholds performs better, and this is emphasized by the tracking error in Figure 5.5.

The importance of the adaptation of the thresholds is emphasized by Figure 5.6 and Figure 5.7 which report, respectively, the evolution of the negative thresholds and the positive thresholds of the SO. It is interesting to see that the estimated thresholds tend towards their real values, which obviously correspond to the optimal values.

The second scenario duplicates the analysis of the first in the case that the alteration of the thresholds is big, i.e. $[w, p] = 0.8[w_0, p_0]$, $r = r_0$ and $a = 0.5a_0$. Figure 5.8 shows a zoom on the tracking performance of the two schemes. The controller without thresholds performs worse than before. This can be appreciated by the tracking error plot of Figure 5.9.

As before, the adaptation of the thresholds shown in Figure 5.10 and Figure 5.11 demonstrates to be effective for the improvement of the tracking performance. The control-loop without the adaptation of the thresholds exhibits a bigger MFAE which affects the tracking result.

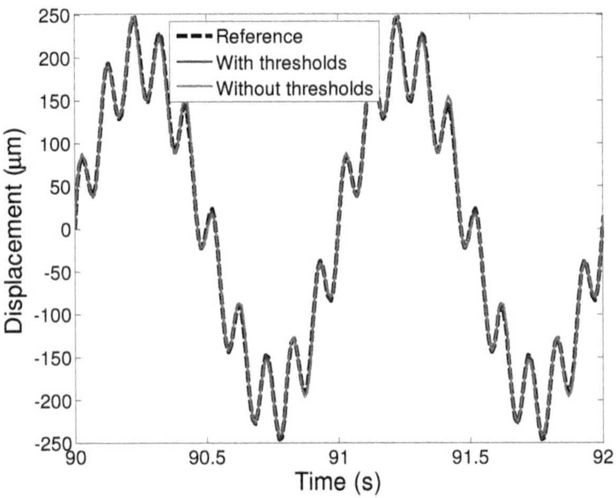

Figure 5.4 – Tracking of a sum of sinusoids (1..10 Hz) and comparison between an adaptive controller which adapts the thresholds and one which does not

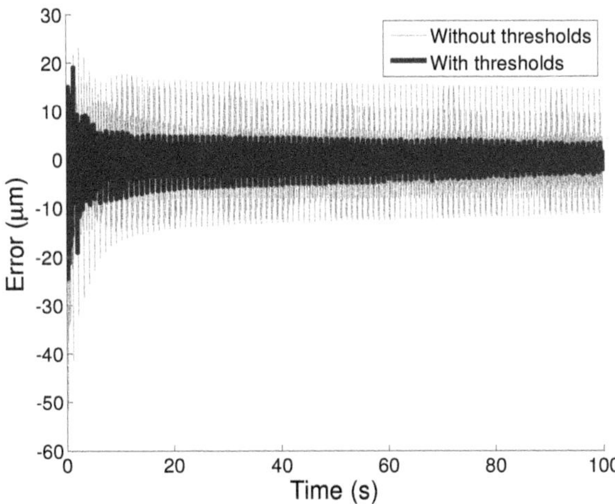

Figure 5.5 - Tracking error and comparison between an adaptive controller which adapts the thresholds and one which does not

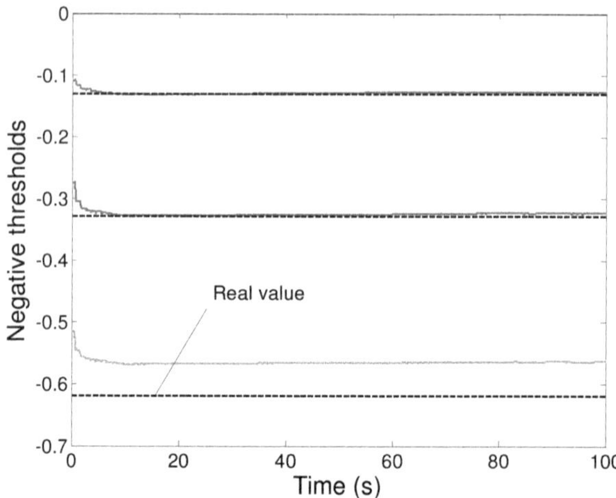

Figure 5.6 – Adaptation of the negative thresholds ($a_l \leq 0$) of the SO and their respective real values (dashed)

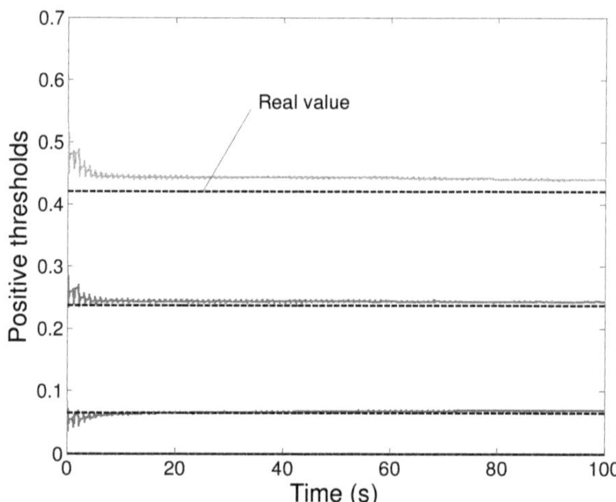

Figure 5.7 - Adaptation of the positive thresholds ($a_l \geq 0$) of the SO and their respective real values (dashed)

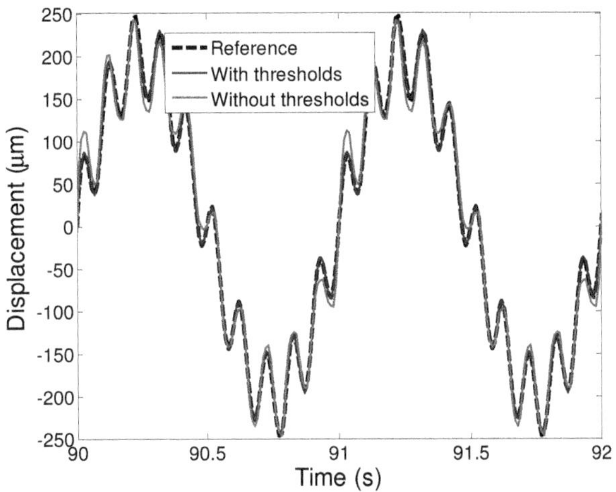

Figure 5.8 - Tracking of a sum of sinus (1..10 Hz) and comparison between an adaptive controller which adapts the thresholds and one which does not

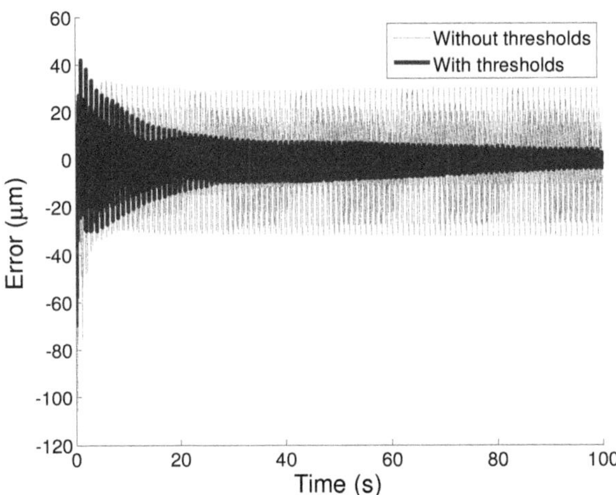

Figure 5.9 - Tracking error and comparison between an adaptive controller which adapts the thresholds and one which does not

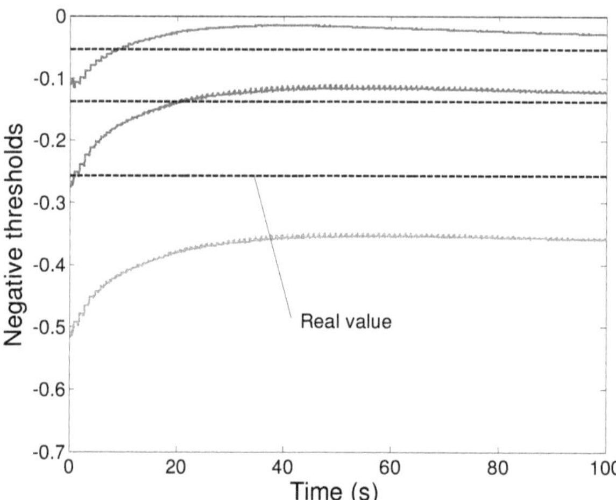

Figure 5.10 – Adaptation of the negative thresholds ($a_l \leq 0$) of the SO and their respective real values (dashed)

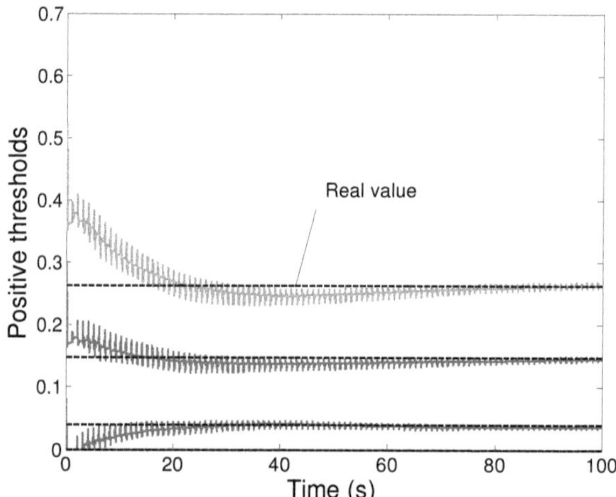

Figure 5.11 - Adaptation of the positive thresholds ($a_l \geq 0$) of the SO and their respective real values (dashed)

5.5. Experimental results

The experimental results that follow are performed on the MSM prototype of Figure 2.4. The plant dynamics have been considered linear and of the first order, as detailed in section 5.4. The adaptation of the thresholds has been turned off by choosing $\gamma_{\Delta}=0$, since in this particular case it was not improving the tracking results. The main goal of these experiments is to validate the overall adaptive framework, with a special emphasis on the adaptation of the hysteresis compensator to deal with temperature disturbances.

In a first step, the hysteresis of the prototype has been identified off-line by elaborating the data acquired at very low frequencies. The obtained hysteresis is already reported in Figure 4.7. The identification provides the initial estimates of the hysteresis parameters. The compensator has been generated with the MPIM→IMPIM mappings given in section 3.3. The plant dynamics have been assumed to be linear and of the first order, but unknown. The control approach is the one discussed in section 5.4, with two minor modifications: 1) instead of the tracking error e_T, the filtered tracking error e_F is used, which contains an integral term, $e_F = K_P e_T + K_I \int_0^t e_T \mathrm{d}\tau$; 2) the presence of control saturation has been taken into account with a simple strategy presented in [81]. Since these modifications do not change the overall framework, they are not discussed here. The interested reader can refer to [80] for more details. The presence of the integral term in e_F allows handling references that are constant as well as references that are not.

We consider three experiments. The first requires the tracking of a sum of sinusoids between 0.1 Hz and 1 Hz. The results of the adaptive controller are summarized in Figure 5.12. The tracking error is within $\pm 3\ \mu\mathrm{m}$.

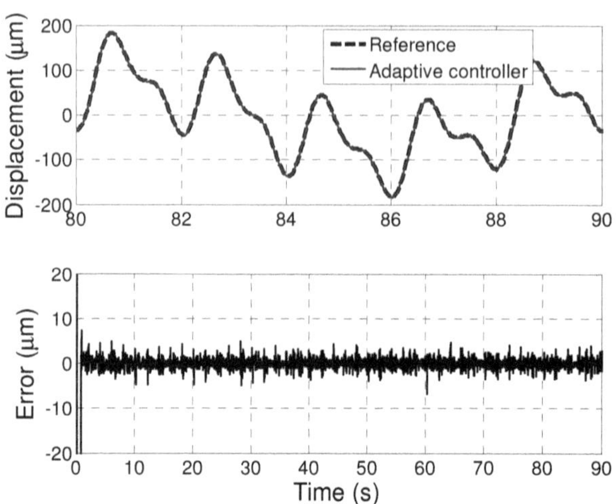

Figure 5.12 – Tracking of a sum of sinus (0.1 Hz..1 Hz)

The second experiment requires the tracking of filtered steps. This signal contains a constant part for which the presence of the integral action is beneficial. However, it also entails a non-constant part. For this reason, this kind of signal is suitable for a comparison between the adaptive controller and a standard PI controller. The comparison is done in two scenarios. In the first, the steps are filtered with a first order linear filter that has a pole at -20 rad/s. The tracking offered by the PI controller, tuned with a trial-and-error procedure, is reported in Figure 5.13. The error is within ± 50 μm. It can be seen that in the small periods where the reference is almost constant the PI controller performs quite well. The main weakness is in the rising and falling parts of the signal, where hysteresis plays an important role. The PI is not able to handle the hysteretic nonlinearity in a satisfactory way. Figure 5.14 (top) shows a zoom of the tracking offered by the adaptive controller, on a time period which is the same as for Figure 5.13 (top). The tracking error, in Figure 5.14 (bottom), is within ± 7 μm.

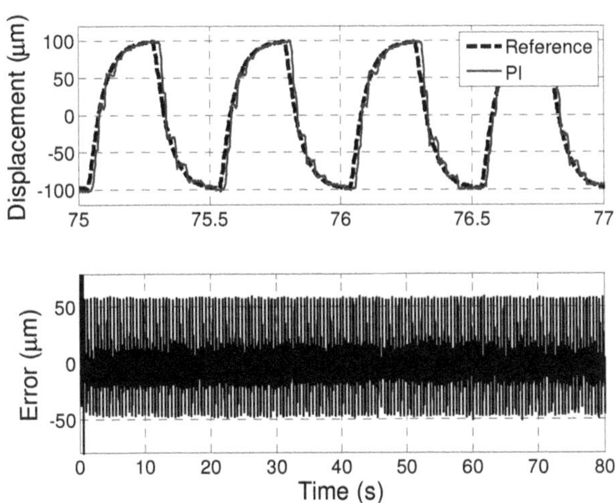

Figure 5.13 – Tracking of filtered steps with a PI controller

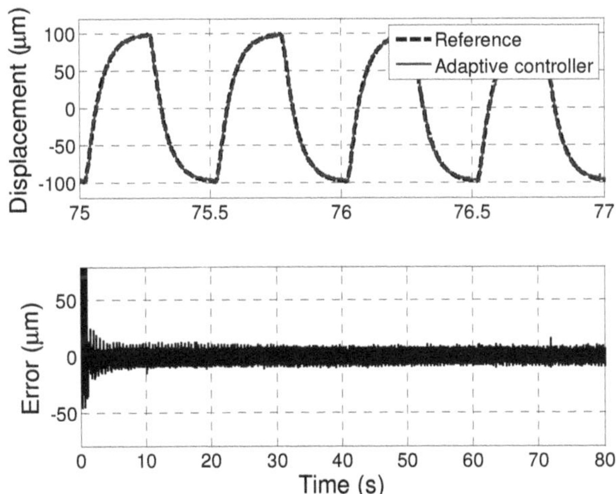

Figure 5.14 - Tracking of filtered steps with the adaptive controller

The second scenario for the comparison between the PI and the adaptive controller requires the tracking of steps filtered by a first order filter with a pole at -50 rad/s. Figure 5.15 shows the performance of the PI controller, whereas Figure 5.16 shows the performance of the adaptive controller. Both controllers have the same tuning as in the previous case. Once again, the adaptive controller exhibits a good error performance of ± 7 μm. The PI controller offers a tracking error within ± 50 μm. However, it can be noted that the PI controller performs well when the filtered steps are almost constant.

The third experiment evaluates the adaptive control approach in the challenging case where the actuator is subjected to externally-induced temperature variations. The tracking is shown in Figure 5.17. The reference is the same sum of sinusoids presented in Figure 5.12. The experiment is done as follows: the actuator is initialized, then it is heated with a heating gun between the first vertical line and the second vertical line of Figure 5.17 (bottom). After, the actuator is cooled down by a fan. It can be seen from Figure 5.17 (bottom) that the tracking error is kept almost constant, despite the thermally-induced variations of the hysteresis characteristic. Thus, the adaptive controller can ensure a tracking performance, even in the presence of *unmeasured* thermal disturbances. To compensate those disturbances, the adaptive MPIM contained in the adaptive controller adapts its parameters (recall that only w, p and the offset σ are adapted). The adaptation is shown in Figure 5.18. During heating, the offset experiences a big increase. This is coherent with what represented in Figure 2.11: when temperature increases, the hysteresis characteristic moves towards the upper side of the current-displacement plane. Coherently, the offset estimation decreases when the actuator is cooled down. The adaptation can also be recognized in the control action plot of Figure 5.19 (in the case of MSM actuators, the control action u is the current int the field-generating coils). The current shifts when temperature increases or decreases.

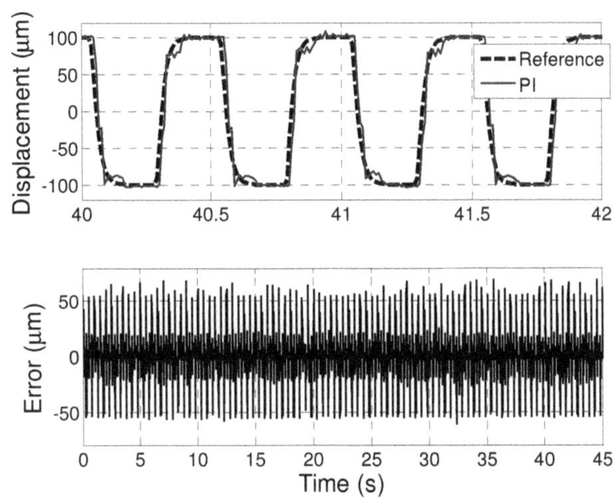

Figure 5.15 - Tracking of filtered steps with a PI controller

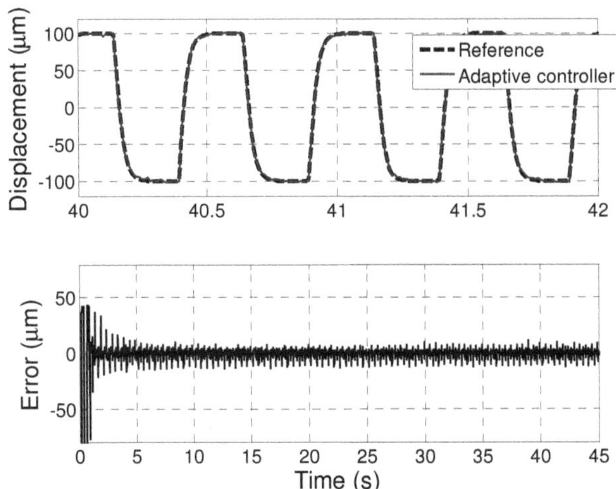

Figure 5.16 - Tracking of filtered steps with the adaptive controller

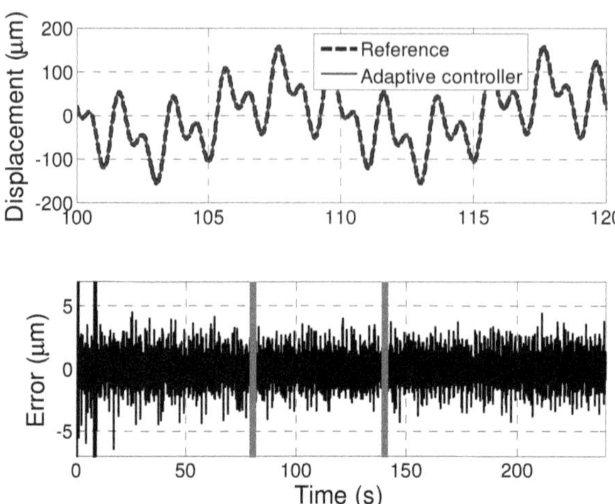

Figure 5.17 – Tracking of a sum of sinusoids with temperature variations

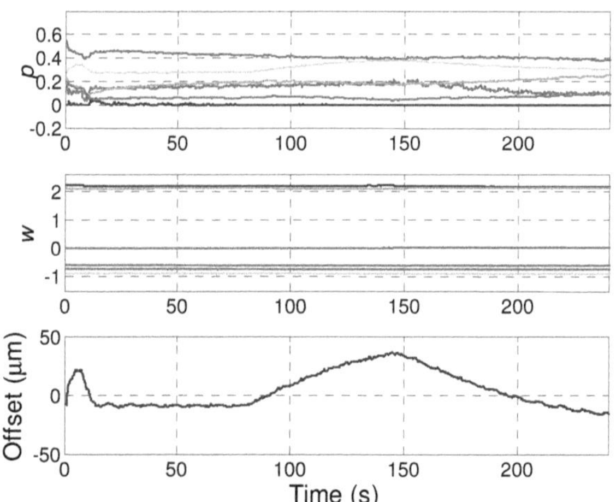

Figure 5.18 – Adaptation of the parameters

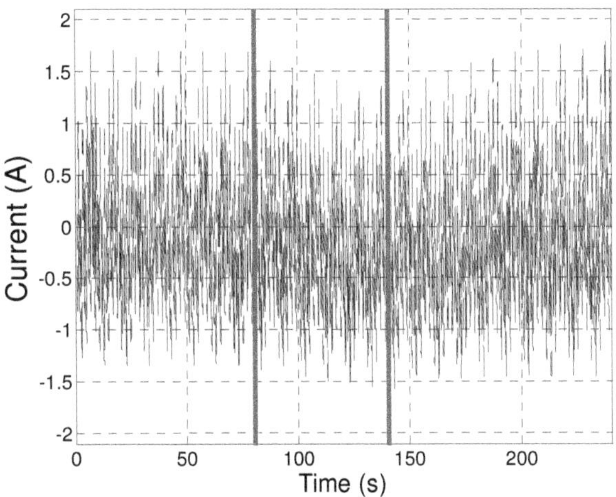

Figure 5.19 – Behavior of the control action (current u) during heating and cooling

5.5.1. Discussion

The adaptive control approach detailed in the previous sections has been developed because of the particular nonlinear behavior of the hysteresis which arises in MSM actuators. The influence of temperature on the hysteresis has been the object of section 2.3.2, where it is shown that temperature modifies the shape and the offset of the hysteretic phenomenon. For this reason, the adaptive control-loop represented in Figure 5.3 traces the thermally-induced variations of the hysteresis by means of an adaptive MPIM, and inverts the hysteresis by means of a IMPIM to compensate for its effects. The dynamics of the plant is taken into consideration by using a feedback linearization strategy based on the adaptive approximation of the unknown terms which determine the evolution of the state.

From the experimental viewpoint, the control approach presented in section 5.3 has two drawbacks. The first concerns the necessity to measure the complete state vector of the plant in order to calculate the tracking error $\tilde{x}$. This is not always possible. For example, in positioning systems like the one that has been analyzed it is often possible to measure only the displacement variable and not the velocity and/or the acceleration. Several methods can be exploited to overcome the mentioned limitation: the state vector can be estimated by adaptive observers. The adoption of state observers can be the focus of future research.

The second drawback concerns the number of variables which are adapted on-line. The hysteresis parameters ρ, ω, Δ can contain many variables, depending on the order of the MPIM. In some unfavorable but realistic situations, the presence of measurement noise can make the adaptation of some parameters practically useless. In the experimental set-up where the experiments have been performed, the displacement sensor showed a measurement noise of about $\pm 3\ \mu\text{m}$, so it has been not possible to further reduce the tracking error by adapting also the thresholds (note that, however, the used thresholds were starting from good initial values offered by the off-line identification of the hysteresis).

Moreover, the number of variables to be adapted makes the choice of their respective adaptive gains an extremely hard task. The experiments of Figure 5.12-Figure 5.19 have been performed without changing the adaptive gains of the adaptive controller, so that the tuning was the same for all the references. A better tuning of the gains for each reference can lead to better tracking results for that specific reference. The determination of the best adaptive gains is an open issue in the adaptive control literature. A step towards a solution has been done by [72]. In this paper, the adaptive gains of the hysteresis compensator are assumed to be much smaller than the adaptive gains of the linear controller. It is proved that the estimated hysteresis parameters converge to their best value, thus ensuring the maximum benefit of the hysteresis compensation. The extension of this analysis to the case of nonlinear dynamics can also be the target for future studies.

There is an important issue which concerns the control of MSM actuators. Refer now to Figure 2.11: when temperature changes, the actuator changes the displacement range. In other words, a specific reference trajectory can be feasible at certain temperatures but not feasible at other ones. The amount of current certainly depends on the particular reference that has to be tracked. Thus, the reference trajectory must be carefully designed based on the temperature range that the actuator experiences during operation. This is emphasized by Figure 5.19: when temperature increases, the current shifts to smaller values, but a further shifting will require a current that is below the saturation limits at $-2\ \text{A}$. The fact that the required current is not in the feasible range means that the required trajectory is not feasible anymore at that temperature value.

Finally, it has to be remarked that temperature induced variations on the input-output behavior of MSM actuators are completely unpredictable. They depend on the particular way in which temperature affects the material. The presented adaptive approach has been able to offer a "plug and play" control component, which works independently of the initial conditions of the actuator, and independently of temperature disturbances, assumed that the required reference signals were always feasible.

Conclusions

This thesis addresses the problem of effectively controlling MSM actuators for positioning purposes. In the first part the MSM materials and actuators are presented. Chapter 2 shows that MSM actuators are affected by hysteresis, and that this hysteresis is strongly influenced by temperature.

For this reason, this thesis proposes two control approaches. In the first, a simple standard controller is used to ensure the tracking of constant references. This is the object of Chapter 4. Here, hysteresis is assumed to be time invariant, a situation that is guaranteed when the MSM actuator design is robust with respect to thermal disturbances, or when the MSM element is insensitive to temperature variations which are bounded in some ranges. The proposed approach extends the available literature since it reduces the problem of controlling a hysteretic system to the stability problem of a linear perturbed system. The effectiveness of the approach is demonstrated by several experimental results. In particular, since the resulting stability problem involves Linear-Matrix-Inequality calculations, several common methods can be exploited to design the standard controller with respect to some required performance (e.g., robust pole placement or robust optimal control).

The second control approach, detailed in Chapter 5, considers a very general scenario: the hysteresis and the dynamics of the MSM actuator are assumed to be unknown, and the dynamics are assumed to be nonlinear. The approach develops an adaptive control strategy which makes use of an adaptive hysteresis compensator to compensate for the hysteresis nonlinearity, and of adaptive approximation techniques to cancel the nonlinear dynamics and linearize the plant. The adaptive hysteresis compensator is the main component of the approach: thanks to the adaptation laws of its parameters, it is able to trace the thermally-induced variations of the hysteresis and compensate for them. The effectiveness of the adaptive control approach is investigated by several experiments. In particular, one of them is performed by disturbing the actuator operation by inducing temperature variations. The adaptive control approach is able to guarantee a constant tracking performance despite of the disturbances.

There are several possibilities for further research about control, and some of them are listed in details at the end of Chapter 4 and Chapter 5. However, a necessary step is the definition of design rules for MSM actuators. Some new concepts have already been presented, and they appear extremely interesting from the application viewpoint but also from the control viewpoint. On one hand, the control approaches proposed in this thesis can be used for several kinds of actuators (and not only with MSM materials). On the other hand, each actuator has its particular features, which require the control approach to be modified. For instance, the push-push actuator presented in Chapter 2 can exploit the twinning stress as an advantage for energy savings. Can the control approach of Chapter 4 ensure an optimal exploitation of that feature? This is definitely a topic of future research. But this also requires the PID control approach to be further developed from the theoretical viewpoint.

The adaptive control of Chapter 5, instead, has one important drawback that should be addressed, and it concerns the choice of the adaptive gains. Although the approach is able to handle a wide class of completely unknown dynamic hysteretic systems, the effort required to tune the adaptive gains requires some time. We have experienced that an *ad hoc* tuning of the adaptive gains for each kind of reference signal (sinusoidal wave, sequence of steps, triangular wave…) was able to offer an improved tracking performance for that specific reference. Future research could focus on the determination of the optimal gains for a fixed reference.

To conclude, we believe that this thesis gives to the interested reader some good ideas about how to design a control system for MSM actuators in positioning applications. Considering the advances in the production of new varieties of

MSMAs, it can be foreseen that positioning systems based on these materials will experience remarkable improvements in the forthcoming years.

References

[1] A. Sozinov, A. A. Likhachev, and K. Ullakko, "Crystal structures and magnetic anisotropy properties of Ni-Mn-Ga martensitic phases with giant magnetic-field-induced strain," *IEEE Transactions on Magnetics*, vol. 38, no. 5, pp. 2814–2816, Sep. 2002.

[2] S. Faehler, "An introduction to actuation mechanisms of Magnetic Shape Memory Alloys," *ECS Transactions*, vol. 3, no. 25, pp. 155–163, 2007.

[3] A. A. Likhachev, A. Sozinov, and K. Ullakko, "Different modeling concepts of magnetic shape memory and their comparison with some experimental results obtained in NiMnGa," *Materials Science and Engineering A*, vol. 378, no. 1–2, pp. 513–518, 2004.

[4] R. N. Couch, J. Sirohi, and I. Chopra, "Development of a quasi-static model of NiMnGa magnetic shape memory alloy," *Journal of intelligent material*, vol. 18, no. 6, pp. 611–622, 2007.

[5] J. Gauthier, C. Lexcellent, A. Hubert, and J. Abadie, "Modeling rearrangement process of martensite platelets in a magnetic shape memory alloy Ni2MnGa single crystal under magnetic field and (or) stress action," *Journal of Intelligent Material Systems and Structures*, vol. 18, no. 3, pp. 289–299, 2007.

[6] N. Sarawate and M. J. Dapino, "Dynamic sensing behavior of ferromagnetic shape memory NiMnGa," *Smart Materials and Structures*, vol. 18, 2009.

[7] B. Kiefer and D. C. Lagoudas, "Magnetic field-induced martensitic variant reorientation in magnetic shape memory alloys†," *Philosophical Magazine*, vol. 85, no. 33–35, pp. 4289–4329, Nov. 2005.

[8] R. C. Smith, *Smart Material Systems: Model Development*. Philadelphia, USA: Society for Industrial and Applied Mathematics, 2005.

[9] I. Suorsa, "Performance and modeling of magnetic shape memory actuators and sensors," Helsinki University of Technology, 2005.

[10] H. Janocha, *Unkonventionelle Aktoren - Eine Einfuehrung*. Muenchen: Oldenbourg Verlag, 2010.

[11] I. Suorsa, E. Pagounis, and K. Ullakko, "Magnetic shape memory actuator performance," *Journal of Magnetism and Magnetic Materials*, vol. 272, pp. 2029–2030, 2004.

[12] J. E. Huber, N. A. Fleck, and M. F. Ashby, "The selection of mechanical actuators based on performance indexes," *Proc. of the Royal London Society*, vol. 453, pp. 2185–2205, 1997.

[13] A. Boehm, S. Roth, R. Chulist, W. Skrotzki, H. Kunze, W. G. Drossel, and R. Neugebauer, "Development Status of Ni-Mn-Ga Ferromagnetic Shape Memory Polycrystals After Hot Rolling," in *International Conference of New Actuators and Drives*, 2008, p. 742.

[14] M. Poetschke, U. Gaitzsch, and S. Roth, "Preparation of melt textured Ni-Mn-Ga," *Journal of Magnetism and Magnetic Materials*, vol. 316, no. 2, pp. 383–385, 2007.

[15] M. Poetschke, S. Weiss, U. Gaitsch, D. Cong, C. Huerrich, S. Roth, and L. Schultz, "Magnetically resettable 0.16% free strain in polycristalline NiMnGa plates," *Scripta Materialia*, vol. 63, pp. 383–386, 2010.

[16] U. Gaitzsch, J. Romberg, and M. Poetschke, "Stable magnetic-field-induced strain above 1% in polycristalline Ni-Mn-Ga," *Scripta Materialia*, vol. 65, no. 8, pp. 679–682, 2011.

[17] D. Y. Cong, Q. Luo, and S. Roth, "Other alloy with at least a higher austenite-martensite transition temperature," *Journal of Magnetism and Magnetic Materials*, vol. 323, no. 20, pp. 2519–2523, 2011.

[18] J. Lu, W. Qu, and C. Wang, "Sensing Characteristics of Magnetic Shape Memory Materials," in *International Conference on Intelligent Computation Technology and Automation*, 2009, vol. 2, pp. 924–927.

[19] J. M. Stephan, E. Pagounis, M. Laufenberg, O. Paul, P. Ruther, A. M. Reorientation, B. M. Field, and I. Strain, "A Novel Concept for Strain Sensing based on the Ferromagnetic Shape Memory Alloy NiMnGa," *IEEE Sensors Journal*, vol. 11, no. 11, 2011.

[20] K. Kuhnen, H. Janocha, and M. Schommer, "Exploitation of inherent sensor effects in magnetostrictive actuators," in *International Conference on New Actuators and Drives*, 2004, pp. 367–370.

[21] I. Karaman, B. Basaran, H. E. Karaca, A. I. Karsilayan, and Y. I. Chumlyakov, "Energy harvesting using martensite variant reorientation mechanism in a NiMnGa magnetic shape memory alloy," *Applied Physics Letters*, vol. 90, no. 17, Apr. 2007.

[22] B. Holz, L. Riccardi, H. Janocha, and D. Naso, "MSM Actuators: Design Rules and Control Strategies," *Advanced Engineering Materials*, vol. 14, no. 8, pp. 668–681, 2012.

[23] K. Schlueter, B. Holz, and A. Raatz, "Principle design of actuators driven by magnetic shape memory alloys," *Advanced Engineering Materials*, vol. 14, no. 8, pp. 682–686, 2012.

[24] B. Holz and H. Janocha, "MSM actuators - magnetic circuit concepts and operating modes," in *International Conference on New Actuators and Drives*, 2010, pp. 307–310.

[25] H. Schmidt, "Dear magnetische formgdaechtnis-Effekt un seine aktorische Anwendung in der Hydraulik," HTW Saarbruecken, 2010.

[26] B. Holz, L. Riccardi, and H. Janocha, "Compact MSM Actuator – A Concept for Highest Force Exploitation," in *International Conference on New Actuators and Drives*, 2012, pp. 663–666.

[27] J. Gauthier, A. Hubert, J. Abadie, C. Lexcellent, and N. Chaillet, "Multistable actuator based on magnetic shape memory alloys," in *International Conference on New Actuators and Drives*, 2006, pp. 787–790.

[28] L. Riccardi, B. Holz, D. Naso, H. Janocha, M. Laufenberg, and E. Pagounis, "A simulation model for a MSM Push-Push actuator," in *International Conference on Ferromagnetic Shape Memory Alloys*, 2011, pp. 149–150.

[29] L. Riccardi, M. Rosmarino, D. Naso, and H. Janocha, "Position control with a Magnetic Shape Memory push-push actuator," in *International Conference on New Actuators and Drives*, 2012, pp. 671–674.

[30] J. Y. Gauthier, A. Hubert, J. Abadie, N. Chaillet, and C. Lexcellent, "Nonlinear Hamiltonian modelling of magnetic shape memory alloy based actuators," *Sensors & Actuators: A. Physical*, vol. 141, no. 2, pp. 536–547, 2008.

[31] H. Tan and M. H. Elahinia, "A nonlinear model for ferromagnetic shape memory alloy actuators," *Communications in Nonlinear Science and Numerical Simulation*, vol. 13, no. 9, pp. 1917–1928, 2008.

[32] H. Janocha, D. Pesotski, and K. Kuhnen, "FPGA-Based Compensator of Hysteretic Actuator Nonlinearities for Highly Dynamic Applications," *IEEE/ASME Transactions on Mechatronics*, vol. 13, no. 1, pp. 112–116, 2008.

[33] M. Al Janaideh, Y. Feng, S. Rakheja, Y. Tan, and C. Y. Su, "Generalized Prandtl-Ishlinskii Hysteresis: Modeling and Robust Control for Smart Actuators," in *Conference on Decision and Control*, 2009, pp. 7279–7284.

[34] L. Riccardi, G. Ciaccia, D. Naso, and H. Janocha, "Control of MSM actuators for precise positioning," in *International Conference on New Actuators and Drives*, 2010, pp. 787–790.

[35] M. Ruderman and T. Bertram, "On system-oriented modeling and identification of magnetic shape memory (MSM) actuators," in *Mediterranean Conference on Control & Automation (MED)*, 2011, pp. 1134–1139.

[36] F. Ikhouane and J. Rodellar, *Systems with hysteresis - Analysis, identification and control using the Bouc-Wen model.* Wiley InterScience, 2009.

[37] J. Song and A. D. Kiureghian, "Generalized Bouc-Wen model for highly asymmetric hysteresis," *Journal of engineering mechanics*, vol. 132, no. 6, p. 610, 2006.

[38] D. S. Bernstein, "Backlash, bifurcation...and the mysterious origin of hysteresis." [Online]. Available: http://aerospace.engin.umich.edu/people/faculty/bernstein/papers/HystTalkJan022008VWebSite.pdf.

[39] R. V. Iyer and X. Tan, "Control of hysteretic systems through inverse compensation," *IEEE Control Systems Magazine*, vol. 29, no. 1, pp. 83–99, 2009.

[40] X. Tan and O. Bennani, "Fast inverse compensation of Preisach-type hysteresis operators using field-programmable gate arrays," in *American Control Conference*, 2008, pp. 2365–2370.

[41] B. Jayawardhana, H. Logemann, and E. P. Ryan, "PID control of second-order systems with hysteresis," *International Journal of Control*, vol. 81, no. 8, p. 1331, 2008.

[42] K. Kuhnen, "Modeling, Identification and Compensation of Complex Hysteretic Nonlinearities: A Modified Prandtl - Ishlinskii Approach," *European Journal of Control*, vol. 9, no. 4, pp. 407–418, Aug. 2003.

[43] H. K. Khalil, *Nonlinear Systems*, 3rd ed. Prentice Hall, 2002.

[44] A. Padthe, B. Drincic, J. Oh, D. Rizos, S. Fassois, and D. Bernstein, "Duhem modeling of friction-induced hysteresis," *IEEE Control Systems Magazine*, vol. 28, no. 5, pp. 90–107, 2008.

[45] F. Ikhouane and J. Rodellar, "A linear controller for hysteretic systems," *IEEE Transactions on Automatic Control*, vol. 51, no. 2, pp. 340–344, 2006.

[46] H. Logemann, E. P. Ryan, and I. Shvartsman, "Integral control of infinite-dimensional systems in the presence of hysteresis: an input-output approach"," *ESAIM: Control, Optimization and Calculus of Variations*, vol. 13, pp. 458–483, 2007.

[47] S. Valadkhan, K. Morris, and A. Khajepour, "Robust PI control of hysteretic systems," in *Conference on Decision and Control*, 2008, pp. 3787–3792.

[48] L. Riccardi, D. Naso, B. Turchiano, H. Janocha, and K. Schlueter, "PID control of linear systems with an input hysteresis described by Prandtl-Ishlinskii models," in *Conference on Decision and Control*, 2012, p. (to appear).

[49] L. Riccardi, D. Naso, B. Turchiano, H. Janocha, and D. K. Palagachev, "On PID Control of Dynamic Systems With Hysteresis Using Prandtl-Ishlinskii Models," in *American Control Conference*, 2012, pp. 1670–1675.

[50] S. Boyd, L. El Ghaoui, E. Feron, and V. Balakrishnan, *Linear matrix inequalities in system and control theory*, vol. 15. Society for Industrial Mathematics, 1994.

[51] F. Blanchini and S. Miani, *Set-theoretic methods in control*. Boston: Birkhauser, 2008, p. 488.

[52] A. Cavallo, C. Natale, S. Pirozzi, and C. Visone, "Limit Cycles in Control Systems Employing Smart Actuators With Hysteresis," *IEEE/ASME Transactions on Mechatronics*, vol. 10, no. 2, pp. 172–180, Apr. 2005.

[53] C.-A. Jiang, M.-C. Deng, and A. Inoue, "Robust stability of nonlinear plants with a non-symmetric Prandtl-Ishlinskii hysteresis model," *International Journal of Automation and Computing*, vol. 7, no. 2, pp. 213–218, May 2010.

[54] R. B. Gorbet and K. a. Morris, "Generalized dissipation in hysteretic systems," in *Conference on Decision and Control*, 1998, pp. 4133–4138.

[55] R. B. Gorbet, K. a. Morris, and D. W. L. Wang, "Passivity-based stability and control of hysteresis in smart actuators," *IEEE Transactions on Control Systems Technology*, vol. 9, no. 1, pp. 5–16, 2002.

[56] A. Cavallo, C. Natale, S. Pirozzi, and C. Visone, "Effects of hysteresis compensation in feedback control systems," *IEEE Transactions on Magnetics*, vol. 39, no. 3, pp. 1389–1392, Nov. 2003.

[57] L. Riccardi, D. Naso, B. Turchiano, and H. Janocha, "A precise positioning actuator based on feedback-controlled Magnetic Shape Memory Alloys," *Mechatronics*, vol. 22, no. 5, pp. 568–576, 2012.

[58] P. Ge and M. Jouaneh, "Tracking control of a piezoceramic actuator," *IEEE Transactions on Control Systems Technology*, vol. 4, no. 3, pp. 209–216, 1996.

[59] J. Shen, W. Jywe, H. Chiang, and Y. Shu, "Precision tracking control of a piezoelectric-actuated system," *Precision Engineering*, vol. 32, no. 2, pp. 71–78, Apr. 2008.

[60] G. Song, J. Zhao, X. Zhou, and J. A. D. Abreu-Garcia, "Tracking Control of a Piezoceramic Actuator With Hysteresis Compensation Using Inverse Preisach Model," *IEEE/ASME Transactions on Mechatronics*, vol. 10, no. 2, pp. 198–209, 2005.

[61] M. Ruderman and T. Bertram, "Discrete dynamic Preisach model for robust inverse control of hysteresis systems," in *Conference on Decision and Control*, 2010, pp. 3463 – 3468.

[62] J. M. Nealis and R. C. Smith, "Model-Based Robust Control Design for Magnetostrictive Transducers Operating in Hysteretic and Nonlinear Regimes," *IEEE Transactions on Control Systems Technology*, vol. 15, no. 1, pp. 22–39, 2007.

[63] X. Fan and R. Smith, "Model-based L1 adaptive control of hysteresis in smart materials," in *Conference on Decision and Control*, 2008, pp. 3251-3256.

[64] L. Riccardi, D. Naso, B. Turchiano, and H. Janocha, "Robust Adaptive Control of Magnetic Shape Memory Actuators for Precise Positioning," in *American Control Conference*, 2011, pp. 5400 – 5405.

[65] K. Kuhnen, "Compensation of parameter-dependent complex hysteretic actuator nonlinearities in smart material systems," *Journal of Intelligent Material Systems and Structures*, vol. 19, no. 12, pp. 1411–1421, 2008.

[66] L. Riccardi, G. Ciaccia, D. Naso, H. Janocha, and B. Turchiano, "Position Control for a Magnetic Shape Memory Actuator," in *IFAC International Symposium on Mechatronic Systems*, 2010, pp. 478 –485.

[67] G. Tao and P. V. Kokotovic, "Adaptive control of plants with unknown hystereses," *IEEE Transactions on Automatic Control*, vol. 40, no. 2, pp. 200–212, 1995.

[68] D. Pesotski, H. Janocha, and K. Kuhnen, "Adaptive Compensation of Hysteretic and Creep Non-linearities in Solid-state Actuators," *Journal of Intelligent Material Systems and Structures*, vol. 21, no. 14, pp. 1437-1446, 2010.

[69] K. Kuhnen and P. Krejci, "Identification of Linear Error-Models with Projected Dynamical Systems," *Mathematical and Computer Modelling of Dynamical Systems*, vol. 10, no. 1, pp. 59–91, 2004.

[70] J. Fu, W.-F. Xie, and S.-P. Wang, "Robust adaptive control of a class of nonlinear systems with unknown Prandtl-Ishilinskii-Like hysteresis," in *Conference on Decision and Control*, 2009, pp. 2448–2453.

[71] Q. Wang, C. Su, and Y. Tan, "On the Control of Plants with Hysteresis : Overview and a Prandtl-Ishlinskii Hysteresis Based Control Approach," *Automatica*, vol. 31, no. 1, 2005.

[72] X. Tan and H. K. Khalil, "Control of unknown dynamic hysteretic systems using slow adaptation: Preliminary results," in *American Control Conference*, 2007, pp. 3294–3299.

[73] K. J. Astrom and B. Wittenmark, *Adaptive Control*, 2nd ed. Prentice-Hall Englewood Cliffs, NJ, 2008.

[74] K. Schlueter and L. Riccardi, "Open-loop control of MSM actuators in presence of temperature variations," in *International Conference on Ferromagnetic Shape Memory Alloys*, 2011, pp. 151–152.

[75] K. Schlueter, A. Raatz, and L. Riccardi, "An Open-Loop Control Approach for Magnetic Shape Memory Actuators Considering Temperature Variations," *Advances in Science and Technology*, vol. 78, pp. 119–124, 2013.

[76] A. F. Timan, *Theory of approximation of functions of a real variable.* Dover, 1993, p. 644.

[77] J. A. Farrell and M. M. Polycarpou, *Adaptive Approximation Based Control.* Wiley InterScience, 2006, p. 440.

[78] J. Y. Choi and J. A. Farrell, "Nonlinear Adaptive Control Using Networks of Piecewise Linear Approximators," *IEEE Transactions on Neural Networks*, vol. 11, no. 2, pp. 390–401, 2000.

[79] L. Riccardi, D. Naso, B. Turchiano, and H. Janocha, "Adaptive Modified Prandtl-Ishlinskii Model for Compensation of Hysteretic Nonlinearities in

Magnetic Shape Memory Actuators," in *International Conference on Industrial Electronics*, 2011, pp. 56 – 61.

[80] L. Riccardi, D. Naso, B. Turchiano, and H. Janocha, "Adaptive Approximation-based Control of Unconventional Actuators," in *Conference on Decision and Control and European Control Conference*, 2011, pp. 958–963.

[81] M. Polycarpou, J. Farrell, and M. Sharma, "On-line approximation control of uncertain nonlinear systems: issues with control input saturation," in *American Control Conference*, 2003, pp. 543–548.

www.ingramcontent.com/pod-product-compliance
Ingram Content Group UK Ltd.
Pitfield, Milton Keynes, MK11 3LW, UK
UKHW021051270726
13967UKWH00012B/574

9 781291 073195